Problems and Methods for Lithospheric Exploration

ETTORE MAJORANA INTERNATIONAL SCIENCE SERIES
Series Editor:
Antonino Zichichi
European Physical Society
Geneva, Switzerland

(PHYSICAL SCIENCES)

Recent volumes in the series:

A Continuation Order Plan is available for this series. A continuation order will bring delivery of ·
each new volume immediately upon publication. Volumes are billed only upon actual shipment.
For further information please contact the publisher.

Problems and Methods for Lithospheric Exploration

Edited by

Roberto Cassinis

University of Milan
Milan, Italy

Plenum Press • New York and London

Library of Congress Cataloging in Publication Data

Main entry under title:

Problems and methods for lithospheric exploration.

(Ettore Majorana international science series; v. 19)
"Proceedings of the fourth course of the School of Applied Geophysics on the in-
tegration of geophysical and geological data, held March 16, 1982, at the Ettore
Majorana Center, in Erice, Sicily, Italy"—T.p. verso.
Includes bibliographical references and index.
1. Geology, Structural—Congresses. 2. Geophysics—Congresses. 3. Seismology—
Congresses. I. Cassinis, R. (Roberto) II. International School of Applied Geophysics.
III. Series.
QE601.P715 1984   551.8.   84-9817
ISBN-13: 978-1-4612-9451-1   e-ISBN-13: 978-1-4613-2373-0
DOI: 10.1007/978-1-4613-2373-0

Proceedings of the Fourth Course of the School of Applied Geophysics on the
Integration of Geophysical and Geological Data, held March 16, 1982, at the Ettore
Majorana Center, in Erice, Sicily, Italy

© 1984 Plenum Press, New York
Softcover reprint of the hardcover 1st edition 1984
A Division of Plenum Publishing Corporation
233 Spring Street, New York, N.Y. 10013

PREFACE

This volume contains several contributions among those held at the
fourth Course of the International School of Applied Geophysics,
Erice, March 15-24, 1982.

The content has been arranged according to the three main topics
covered during the course: the goedynamic models, the contribution
of geophysics to their construction and the optimization and
constrains of the exploration methods.

At the end of the volume, four short notes are published, written
by participants to the course and related to the general subject.
It is regrettable that not all the lectures and discussions are
available for publication in this volume. However, it is hoped that
the papers presented here will stimulate further discussions and
suggest new topics for future meetings.

                              R. Cassinis

CONTENTS

INTRODUCTORY NOTE

Roberto Cassinis

Institute of Geophysics
University of Milan
Via L. Cicognara 7, 20129 Milano

The basic discoveries made in the sixties by geophysical methods gave origin to the hypothesis of plate tectonics. Then, the increased intensity of the surveys revealed details that could not be explained by the general scheme. This is especially true for the continental lithosphere, where the investigation has proved to be more difficult than in oceanic areas, owing mainly to the occurrence of stronger lateral variations. This type of development is common to all branches of science.

Geology also builds up its models on schematic general hypotheses; when new data seem to contradict the model, they are considered, at least during a first phase, as "disturbing factors" and every effort is made to accommodate the evidence to the scheme instead of using it as proof against the hypothesis. The abundance of new information gathered has now started a second phase of knowledge: the evidence brought by independent information must be used to correct the model and to discard the over-generalized assumptions.

The improvement of geophysical methods, especially during the last decade, has revealed subtleties that may not have been appreciated in the previous research. Even more fundamental progress has been the use of integrated data by structural geologists having a deep knowledge both of the potential and of the constraints of the geophysical methods as well as of the other tools helping the synoptic evaluation of an area, e.g., the images taken from the orbiting Earth satellites. On the other hand one must be aware of the limitations of the exploration methods and try to optimize their use by the best possible statement on the objectives.

The IV Course of the School of Applied Geophysics at the Majorana Center was originally intended to deal with the particular models of "collisional" zones as well as with the problems involved in their exploration. During the Course, discussions arose on the term "collisional" itself; moreover, the subjects treated showed that the problems of the exploration of the lithosphere must be considered on a global basis, the comparison between the types of crustal and upper mantle structure being essential for a correct application of geophysical methods. For this reason, the title of this volume has been modified.

Of course, the models discussed and the results especially deal with transitional zones that sometimes can be defined as "collisional". These areas, the unstable or mobile belts of the world, are those where the internal life of our Planet is most clearly shown. The importance to investigate these areas and to understand their dynamic processes has been stressed recently by the evidence that goals of immense economic value are involved both in a positive and in a negative sense (like the survey for new mineral resources or the mitigation of earthquake hazards).

Three fundamental topics have been treated in the lectures: geodynamic models, geophysical data as well as the iterative process among them; and, last but not least, the available methods for the exploration, their constraints, their optimization and their future developments.

GEODYNAMIC MODELS AND GEOPHYSICAL RESULTS

The Alpine orogenesys and the Mediterranean region have been the preferred topics (Giese, Sholpo, Wezel). The comparison between the Caucasus and the Alps seems particularly stimulating. The Mediterranean is one of the regions of the world where the schematism of the plate tectonics finds more difficulties to fit the complexity of the structure. Probably, a deeper knowledge of the present structure of this area will give a major input to the necessary refinement of the "global tectonics".

The transition from the oceanic to the marginal lithosphere and then to the continental lithosphere as well as the accompanying lateral variations of the asthenosphere have been described by Kosmiskaya and Nolet and illustrated by the interpretation of long lithospheric profiles (active and passive seismology). The results do not fully comply with the essential postulate of global tectonics, viz the spreading of the oceanic floor as the only mechanism responsible for the evolution and for the horizontal movement of continental masses.

Cassinis, Giese and Eva presented studies on particular problems. The seismogenesis in Italy is complicated by the complexity of the geological history and by the consequent change of stress direction at different levels in the crust. Integration of methods is compulsory while dealing with this type of survey as well as for the interpretation of the geothermal anomaly in Tuscany, the largest in Europe, where resistivity and velocity measurements help to make a choice between the possible thermal models. Seismicity of a particular portion of the Indonesian arc has been investigated, showing that even the typical island arcs are divided into compartments of different geophysical character by transverse shifts.

GEOPHYSICAL METHODS

Seismic exploration has been the method most extensively treated. The lectures on this subject have been started by a review paper by Helbig and Schmoll who examined the potential and the constraints of near vertical reflection seismics, that, at least theoretically, is the only method able to reveal the fine structure of the crust. Phinney continued on this topic describing the field and processing methods employed by COCORP and showing some of the results obtained along the continental profiles. A particular and difficult problem due to the type of the recorded events, is migration. The conclusion is that the features which are most easily seen are layered in some sense (deep sedimentary basins, layered crystalline complexes, low angle thrust faults). The significance of the signals coming from a depth compatible with the "M" transition is still debated.

Kosmiskaya, Scarascia and Giese discussed the potential and constraints of the more widely used technique for deep exploration, the wide or critical angle refraction, also named "deep seismic soundings". Especially in mobile, orogenic belts, problems arise on the reliability of data: refinements of field and interpretation techniques are needed. In spite of their lower resolution and their sensitiveness to lateral variations, the deep seismic soundings still are the only means to obtain the velocity distribution beneath the sedimentary cover.

Nolet described the advanced tools of passive seismology, using both surface and body waves of teleseismic events, to obtain a detailed upper mantle velocity model to a depth of 700 km under a region of the size of western Europe.

Alfano dealt with the electromagnetic methods. Active geoelectrical methods, using suitable techniques (dipole-dipole) are preferable to study the resistivity distribution in the upper crust to about 10-15 km, especially where a complicated structure exists. For larger depths natural fields must be measured. The factors limiting the reliability of the latter type of investigation are particularly large; however magnetotellurics and geomagnetic soundings are the only techniques able to give some information on the resistivity distribution in the lower crust and in the mantle.

Helbig chaired a round table on all methods available for the exploration of the lithosphere. The state-of-the-art, the constraints and the future development of the geophysical techniques have been discussed as well as the merging of different data sets.

Finally, four short contributions held by participants were presented, that dealt with problems or techniques particularly related to the study of transitional areas.

GLOBAL SEISMOLOGY AND THE INVESTIGATION

OF DEEP CONTINENTAL STRUCTURE

Guust Nolet

Vening Meinesz Laboratory
P.O. Box 80.021
3508 TA,Utrecht

INTRODUCTION

It is generally accepted wisdom that the highest resolving power for the investigation of the crust and uppermost mantle is obtained through the use of seismic profiling with controlled, high-frequency sources. If we interpret "resolution" as the ability of elastic waves to be scattered by thin layers or small changes in the properties of the rock, there is no reason to question this statement. But to those who have experience with high-frequency seismology in areas of complicated geological structure (usually the interesting areas) it is also evident that this resolution may be so large, that one does not see the wood for the trees: different interpretations of the results of such surveys sometimes lead to seriously conflicting models. Misidentification of phases is usually the cause of conflicts. For instance, we may wrongly interpret a multiple in the crust as a reflection from the mantle. Misidentification is a plague for which no robust editing method can provide the cure, since the error is by nature systematic.

Once we accept a certain identification, depths to discontinuities may be well-constrained, and velocities accurately determined. Yet our resolving power is very bad, since widely different models may be constructed, based on different identifications. What is needed is independent evidence, if possible unambiguous, against which to test hypothetical model interpretations. Teleseismic data can provide such evidence. They do so at low cost and with varying degrees of effort involved.

It is true that there exist ambiguities in the identification of many teleseismic phases as well. However, the propagation of P and S body waves from epicentres at large distance can usually be modelled adequately with 1-dimensional layered media. Often we may assume the incoming wavefront to be plane and our Earth model to be flat (usually these approximations are valid if our structure is not deeper than about 150 km). In this case the simple method of matrix propagators can be used to calculate the response of layered systems, that is to calculate realistic synthetic seismograms, to see if phase identifications make any sense. The theory of propagator matrices can be found in any of the recent textbooks on seismology (Pilant, 1979; Aki and Richards, 1980; Ben Menahem and Singh, 1981; Kennett, 1982), and the reader is referred to one of these for further detail.

In this paper I shall briefly review a number of observational methods that can be used to investigate the crust/lithosphere system at low resolution. I have not tried to compile a complete list of references; citations are limited to those papers that give a practical description of the method and to some recent applications that may provide further references. I shall start with simple time-domain methods. When we consider the interference between converted and reflected waves, as in the spectral ratio method, we shall see that it is worth the trouble to move over to the frequency domain (i.e. use propagator matrices), where we shall finally encounter one of the most useful data for the investigation of crust and lithosphere: the dispersion of surface waves from nearby earthquakes.

TELESEISMIC DELAY TIMES

The easiest, and least ambiguous to observe, is the primary onset on the seismogram, or P-wave. It is the experience of the seismologist that there are stations where this P wave arrives consistently later (or earlier) than expected if the Earth were radially symmetric, no matter where the earthquake is located. Apparently the delay (or gain, which we shall consider a negative delay) is acquired in the region near the station, and contains some information on the crust/lithosphere system there. It is therefore useful to measure these delays. For most existing larger stations this has been done, and delays can be found in the literature (recently by Poupinet, 1979). For temporary or new stations a separate determination of the delay times may be necessary.

Even if one is able to measure the P-onset with a good precision from a short-period seismogram (an accuracy of 0.2 seconds or better can often be realized), one has only measured one end of the interval that is usually called the 'travel time' of the P-wave.

The other end – the origin time of the earthquake – is in general
not available with the same precision. Different methods have been
used to reach the highest possible precision for the measurement
of the travel-time interval. By far the easiest is to use nuclear
explosions and assume that the military officials involved have
very punctual minds and love to detonate their toys at the whole
minute.(e.g. Fairhead and Reeves, 1977). Nuclear tests, however,
are not very common and do not offer a good geographical distribution of sources. Recently it has become popular to do a
statistical analysis on the observed times that are sent to the
International Seismological Centre in Edinburgh (Dziewonski et al.,
1977; Poupinet, 1979), published in monthly Bulletins and stored
on magnetic tape. Since there is anyhow an element of statistics
involved in the determination of station delays (see the following),
the use of statistics on large data sets such as published by the
ISC seems logical, even if the reading of P-onsets is done by
local station operators, so that reading errors are beyond the
control of the investigator himself.

In order to determine the 'delay' one has to compare the
measured travel time with the time given in the Jeffreys-Bullen
tables (1940). For this one has to know the epicentral distance
and the depth of the earthquake. Here, again, there is an error
involved. Fortunately this error may partly cancel the error in
the origin time if these tables themselves are used in the
location of the earthquake, as is done by the ISC. Yet one should
not be overly optimistic about the final accuracy that can be
obtained. During a recent investigation of the East African Rift
(Nolet and Mueller, 1982), I searched through the literature for
station delays in station Addis Ababa. Five different papers gave
different values for the delay, ranging between + 1.16 and + 2.39
seconds, a difference of more than one second !

The theory of delay-time determination is rather simple.
The observed arrival time $\hat{T}_{pq}$ in station p from event q can be
decomposed into a theoretical time T_{pq}, predicted from a standard
Earth model, the delay D_{pq} along the path between p and q, and
a term containing reading and clock errors:

$$\hat{T}_{pq} = T_{pq} + D_{pq} + \epsilon_1 \qquad (1)$$

For one station, and many events, the delays are expected to have
a systematic component R_p due to the delay acquired in the direct
vicinity of the station, and a component which is random if the
earthquakes are well distributed around the globe:

$$D_{pq} = R_p + \epsilon_2 \qquad (2)$$

so that

$$\Delta T_{pq} = \hat{T}_{pq} - T_{pq} = R_p + \varepsilon_1 + \varepsilon_2 \tag{3}$$

A convenient estimator for R_p is found by summing over N events such that the random components ε_1 and ε_2 average out:

$$R_p \approx \frac{1}{N} \sum_{q=1}^{N} \Delta T_{pq} \tag{4}$$

Some care must be taken to exclude anomalous readings (such as misidentifications) and to reduce the influence of source region heterogeneities.

At teleseismic distances, the angle of incidence γ of the P-wave varies little, so that a simplified but approximate interpretation of R_p is possible. If the velocity in a layer of thickness H is changed by an amount ΔV, we find:

$$R_p = - \int_0^H \frac{\Delta V}{V^2 \cos\gamma} \, dz \approx - 0.14 \frac{\Delta V}{V^2} H \tag{5}$$

where the factor 0.14 was calculated for an event at epicentral distance 60°, but may be considered constant to the precision of this approximation for distances between 30° and 90°. Alternatively, if the delay is caused by a change Δz in Moho depth we have (again as a rule of thumb):

$$R_p \approx \Delta z \left(\frac{1}{V_c} - \frac{1}{V_m} \right) \approx 0.026 \, \Delta z \tag{6}$$

where Δz is in km and R_p in seconds. V_c and V_m are the velocities in crust and mantle, respectively and $\gamma \approx 0$ is assumed in the crust. Depending on the local situation (5), (6) or a combination can be used to constrain velocity models. If several stations are available, (6) can be used to convert a trend in R_p to a local trend in Moho depth. In general, R_p is too large to be caused by crustal thickness variations only.

Instead of inverting for crust or mantle structure one may consider R_p as an independent datum and correlate this with other geophysical quantities. Poupinet (1979) finds the following relationship between R_p (in seconds) and the basement age A (in 10^9 years) for continental regions:

$$R_p \approx 1.1 - 1.4 A^{\frac{1}{2}} \tag{7}$$

With assumptions on the temperature dependence of the seismic velocity, and the depth to the asthenosphere in Western Europe, Poupinet modifies (7) to yield the lithosphere thickness in terms

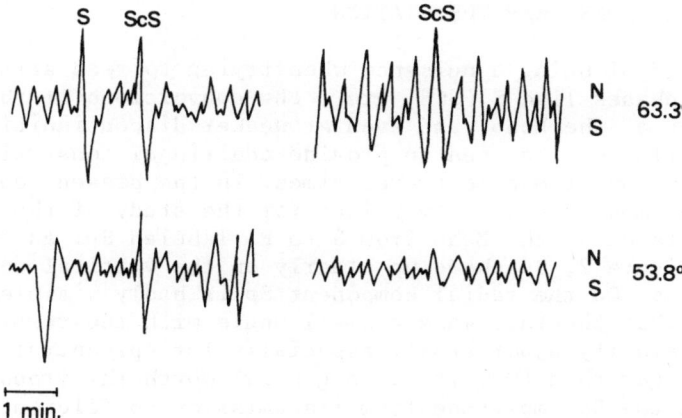

Figure 1   Observations of ScS, and of its first multiple ScS_2 at two epicentral distances. The surface reflection points are beneath a stable continental area (after Sipkin and Jordan, 1976).

of the basement age:

$$H_{lithosphere} \approx 10 + 140A^{\frac{1}{2}} \quad (km) \qquad (8)$$

Fairhead and Reeves (1977) deduce the following relationship between R_p (in seconds) and the Bouguer anomaly B (in mgal) on the African continent:

$$R_p \approx - 1.8 - 0.013B \qquad (9)$$

By combining different rays with partly common ray paths, one may obtain more than one constraint on the structure (Frohlich and Barazangi, 1980), or sample structures other than the near station region (Okal and Anderson, 1975; Sipkin and Jordan, 1976). With later arrivals, such as ScS_2 in Figure 1, care must be taken that the correct waves are picked, and that no bias is introduced by waveform distortion.

WAVE CONVERSIONS NEAR THE STATION

Instead of being a nuisance when trying to read arrivals of the main phases like S, SKS, etc., the conversions of these waves to P waves at the Moho, and even at weaker discontinuities in the upper mantle, can be used to provide additional constraints on crustal and upper mantle travel times. In the present context, one of the most useful conversions for the study of the crust is the conversion at the Moho from S to P, labeled Sp. As can be seen in Figure 2, Sp shows up clearly on the vertical component seismograms. On the radial component Sp is badly visible due to the fact that the rays make a small angle with the vertical in the low-velocity upper crust, especially for epicentral distances longer than $40°$. It is in general worth the trouble to calculate the Sp amplitude from transmission coefficients (Young and Braile, 1976, give a suitable Fortran program). This gives an extra check on the identification of the observed phase. In more complicated crustal models it may be necessary to use the flat Earth-plane wavefront approximation to calculate simple synthetic seismograms with a propagator matrix, as has been done by Kanasewich et al. (1973), to study additional complexities in the S-wave group.

The Sp phase arrives several seconds before the S phase. Jordan and Frazer (1975) find 5 to 6 seconds differences for stations in eastern Canada. The interpretation of the travel time difference between S and Sp is unfortunately not straightforward. For our purpose we may approximate by taking the crust homogeneous with thickness H, and assuming a plane wavefront (i.e. teleseismic event). The travel time difference between S and Sp then becomes

$$\Delta T = H \left[\frac{1}{V_s \cos \gamma_s} - \frac{1}{V_p \cos \gamma_p} \right] \tag{10}$$

(see figure 3).

Thus, if we wish to determine H, we must have information on the P and S velocities, averaged over the crust. Average S velocities can be deduced from Rayleigh or Love wave dispersion if nearby events are at hand. If not enough information is available, Poisson's ratio may be constrained to lie between reasonable values, so that (10) still gives limitations for models resulting from seismic surveys.

10

One more constraint can be obtained from the conversion
if the Sp/S amplitude ratio is measured. Jordan and Frazer (1975)
show that this ratio is very sensitive to the shear wave contrast
at the Moho, much more than to the compressional wave contrast or

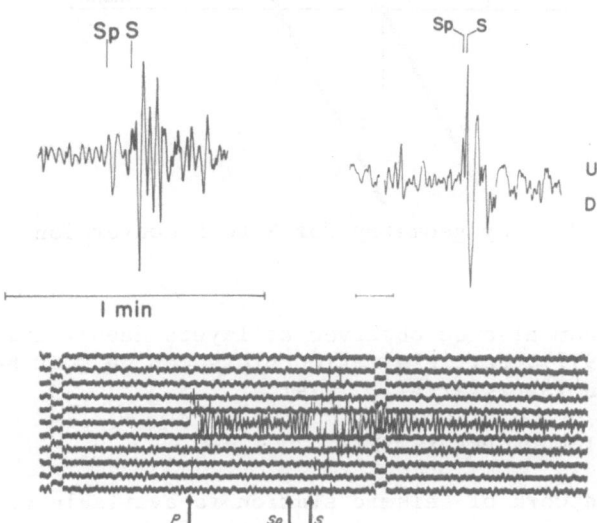

Figure 2   Observations of Sp (vertical components). The upper
            part of the figure shows the short- and long-period
            recording of a conversion beneath station LND in
            Canada (Jordan and Frazer, 1975). Lower seismograms:
            short period recording of Sp in station WEL in
            New Zealand (Smith, 1970).

to reasonable variations in the density. Again, the best way to
proceed is to use the propagator matrix to calculate synthetic
seismograms, applying trial-and-error to match the observed
signal amplitudes and find the S-velocity contrast.

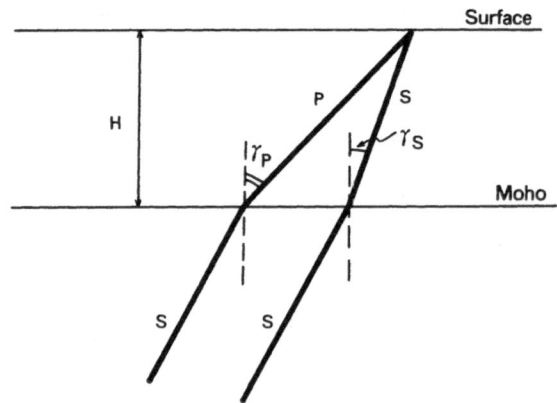

Figure 3   Ray geometry for S to P conversion.

Conversions can also be observed at layers deeper than the
Moho (Vinnik, 1977). Particle motion plots can be most helpful
in identification (Sacks et al., 1979).

3-D INTERPRETATION OF DELAY TIMES

If a dense network of seismic station is available one may
notice different delay times as a function of the location of the
stations for the same event. Since the origin-time error for an
event is now common to all the delays, we can do away with this
source of uncertainty if we measure the relative delays. The
method has been developed for application under the NORSAR and
LASA arrays that have been installed in Norway and Montana (U.S.)
respectively, to monitor the testing of nuclear bombs (Aki et al.,
1976, 1977). Since then, however, it has been applied with networks on widely different scales, ranging from large continental
areas (Romanowicz, 1979, 1980; Menke, 1977; Hirahara, 1977) to
very local structures (Ellsworth and Koyanagi, 1977; Mitchell
et al., 1977; Reasenberg et al., 1980). In general, the method
works best when the network is very dense. Recent advances in
(digital) instrumentation and remote communication make it likely
that the 3-D method will find application in a growing number of
areas in the near future.

In contrast to the averaging over all ray paths that is used
for common interpretation of delay times, the 3-D method retains
the ray path distinction in order to map 3-dimensional velocity
anomalies beneath the sensors. The delays are observed in a

12

number of stations (labeled p) and events (q). Typically, a few
thousand event-station pairs are used. To avoid the influence of
origin-time errors, the average delay over all of the stations is
substracted for each event separately. Theoretical arrival times
are calculated for a reference model that may be homogeneous
(Aki and Lee, 1976), plane-layered (Aki et al., 1977) or more
complicated (Whitcombe, 1982). The reference model is divided
into blocks with a size comparable to the typical sensor distance
at the surface. From the delay times we try to find the velocity
anomaly in each of the blocks. The assumption is that the
lateral heterogeneity is small, such that the delay is caused in
first order by the changed travel time in a block, but that the
change in ray path does only have a second-order influence on the
travel time (we do not perturb the block interfaces). The theori-
tical basis for this is of course Fermat's principle, which states
that these refracted rays follow a minimum time path so that small
changes in the ray path have only a second order effect on the
travel time. The linear relationship between the corrected
travel time delays and the fractional slowness perturbation ΔM_r
in block r is given by:

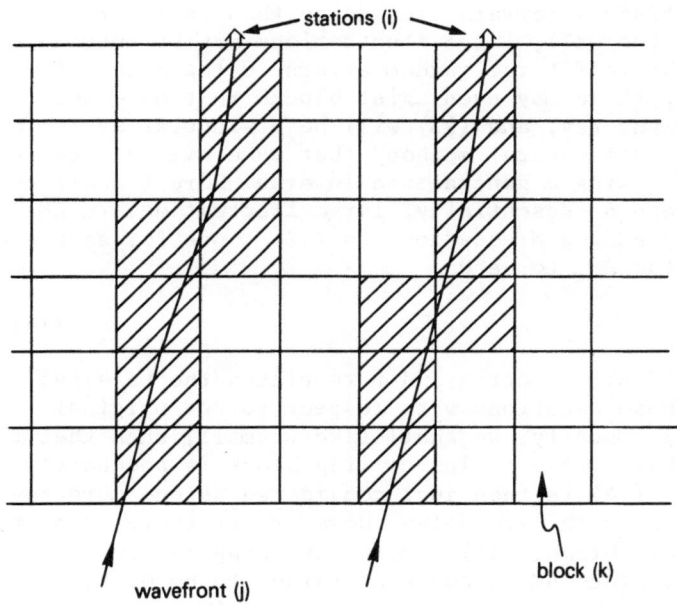

Figure 4   Ray geometry for the 3-D method.

$$(T_{pq}^{obs} - T_{pq}^{cal}) - \overline{(T_{pq}^{obs} - T_{pq}^{cal})} = \sum_r (G_{pqr} - \overline{G_{pqr}})\Delta M_r \qquad (11)$$

where G_{pqr} is the travel time spent by ray (pq) in block r, and the bar denotes the average for each event over all stations. We may write this in matrix notation:

$$\vec{\Delta T} = G \vec{\Delta M} \qquad (12)$$

The well-known least squares solution to this problem is (Franklin, 1968):

$$(G^T G)^{-1} G^T \vec{\Delta T} = \vec{\Delta M} \qquad (13)$$

where T denotes 'transpose'. (13) is a system of linear equations. There are as many linear equations in (13) as there are blocks in the system, say N. Anybody who tries to solve this system with a simple Fortran routine from the local computer library is in for an unpleasant surprise: most probably the solution ΔM_r will fluctuate heavily from block to block and the resulting model can be far away from geological reality. The reason is that some blocks are visited by many rays, but others - notably on the side of the model - by only one or two rays that may have sampled the block for only a very short path near a corner. In these blocks much larger slowness changes are necessary to satisfy a certain time delay than in others. If we try to fit (13) too well, large fluctuations may be introduced into these blocks to fit the random errors in the data. If we are not careful, there may even exist blocks that have not been visited by any ray, and (13) will be a singular system of equations. There are several methods that take away all these worries, such as using a generalized inverse. Here I shall give a very simple method. Essentially, large fluctuations in $\vec{\Delta M}$ can be damped by adding N equations to (12) - one for each block - that constrain the ΔM_r to be 0 :

$$\mu I \vec{\Delta M} = 0 \qquad (14)$$

where I in the identity matrix and μ regulates the relative importance of these equations with respect to the original data system (12). Usually, we shall take μ small, such that the added equation has little influence if a block is adequately sampled by rays. (14) is then just considered as one more ray through the block, with zero delay. However, if there is no ray crossing a certain block, (14) removes the singularity of the original system, since ΔM is now constrained to be 0. For badly constrained blocks the situation is somewhere in between and (14) will act to damp extreme fluctuations. The method is therefore usually named 'damped least squares'. If we combine (12) and (14) the resulting least squares system is in general stable:

$$(G^T G + \mu^2 I)^{-1} \; G^T \; \vec{\Delta T} = \vec{\Delta M} \tag{15}$$

'small' for μ can now be understood to imply that μ^2 is small
with respect to the diagonal elements of $G^T G$. Some experimenting
with μ is generally necessary to find the right degree of damping.
More sophisticated methods also include an analysis of the
resolution of the system (Aki et al., 1977).

THE SPECTRAL RATIO METHOD

Up to now, we have considered travel times. We did show how
converted waves can be sought for and interpreted, and saw the
difficulties associated with multiple reflections and wave
conversions. Reverberations in a layered system such as the crust
often become too complicated to be analysed in terms of separate
arrivals in the time domain, mainly because new arrivals are
masked by the coda of earlier phases. Within the stack of layers,
the wave field becomes rather complicated with a multitude of
refracted and reflected waves interfering constructively or
destructively, depending on their differences in phase.

Fortunately, when calculating the wave transmission through
such a stack of layers, we do not need to keep track of every
possible ray path. The propagator matrix methods incorporate all
(multiple) reflections and refractions. The price one pays for
this is that the reflection- and transmissioncoefficients as a
whole become frequency-dependent, since phases depend on the wavelength, so that the Fourier Transform will turn up at one stage or
another in the process of data analysis. In the time domain, this
dispersive property of the crust will change the waveform shape
of any seismic phase that is transmitted through it from below.
We would of course like to deduce the major properties of the
crust from the change in the form of the incoming wave. But here
we face a problem: since there is no seismometer at the bottom
of the crust, we do not know the shape of the incoming wave. What
can we do about this?

Phinney (1964) devised an ingenious solution to this problem.
His analysis is wholly in the frequency domain, and therefore
involves digitization and Fourier-transformation of the
- suitably windowed - P pulse. The principle of this so called
'spectral ratio method' is simple. When a P wave with an (unknown)
spectrum $S(\omega)$ impinges on a stack of layers such as the crust,
this stack will act as a filter, since the transmission coefficients
are frequency dependent. Phinney makes use of the fact that these
transmission coefficients are different for the vertical and the
horizontal (radial) components of the P-wave.

Thus we can write:

$$u_r(\omega) = S(\omega) \ T_r(\omega \ ; \ p) \tag{16}$$

$$u_z(\omega) = S(\omega) \ T_z(\omega \ ; \ p) \tag{17}$$

where ω is the circle frequency and p is the slowness of the wave. Note that we work in the flat Earth-plane wavefront approximation so that p is assumed the same for all multiply-reflected waves within the crust. By taking the ratio of (16) and (17) we can now, at least in theory, eliminate the unknown spectrum $S(\omega)$ of the incoming wave:

$$R(\omega \ ; \ p) = u_z/u_r = T_z(\omega)/T_r(\omega) \tag{18}$$

The right-hand side of (18) can be calculated for crustal models, and usually trial-and-error inversion is used to find one or more models that satisfy the observed ratio R. As a rule of thumb, the frequency of the first peak in the spectral ratio plot is very close to $1/(2t_s)$, where t_s is the travel time of the shear wave in the crust (see figure 5). At this frequency the denominator in (18) is small because the horizontal component interferes destructively, and the location of the resulting peak can be used as a guide in modelling.

In practical applications of the method, one should pay attention to a number of complications that arise. Much of the earlier literature is devoted to methods to overcome these problems. First of all, one calculates the spectra using a narrow window in the time domain, in order to exclude subsequent phases like pP. It has been shown by Leblanc (1967) that one can approximate this windowed ratio by

$$\text{windowed} \ \frac{u_z}{u_r} = \frac{T_z(\omega) \ \star \ W(\omega)}{T_r(\omega) \ \star \ W(\omega)} \tag{19}$$

where $W(\omega)$ is the spectrum of the time window and $\star$ denotes convolution. Furthermore, instabilities in the estimation of the spectrum can be smoothed by averaging over many earthquakes. Fortunately, the shape of R is only weakly dependent on the slowness p, and Phinney (1964) shows that it is permissable to average over a range of epicentral distances. Kurita (1973) has attempted to extend the method to include mantle reverberations as well. Probably also because digital registration makes the application of the method not very laborious, there seems to be a revival of the spectral ratio method in recent years. One good example is the paper by Bungum et al. (1980), who study variations in crustal thickness in Fennoscandia. Their recommendation is to use at least 5 events for each station and to experiment with different window lengths (1 to 2 minutes) to ensure stability

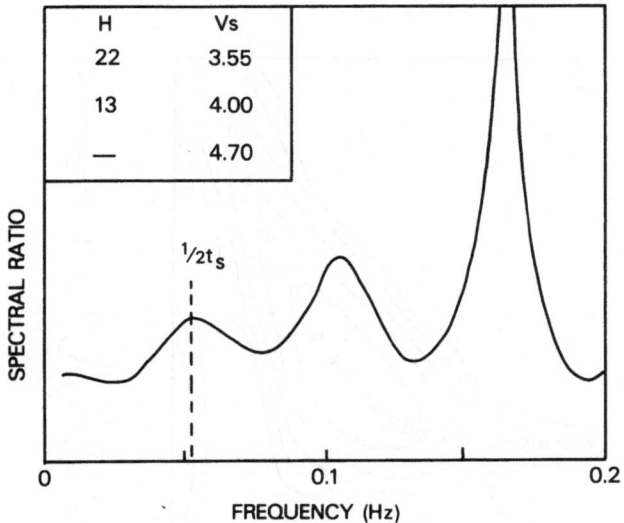

H	Vs
22	3.55
13	4.00
—	4.70

Figure 5   Spectral ratio for a simple crustal model
(after Phinney, 1964).

of the analysis. The method is usually able to resolve 2 or 3
layers in the crust and to determine the Moho depth with an
accuracy of a few kilometers.

SURFACE WAVES

A surface wave can be seen as the combined effect of waves
diffracted at the Earth's free surface and the interference of
many multiple reflections of rays channelling in the crust or
upper mantle. Provided the Earth is locally isotropic and
laterally homogeneous, there are two types of surface waves,
distinguished by the polarisation of their particle movement:
Rayleigh waves (of the P-SV type) and Love waves (SH type).
In analogy to the vibrations of air in an organ pipe, surface
waves have a groundtone and overtones, usually called
'fundamental' and 'higher modes'. In realistic Earth models,
the low frequency components of a surface wave travel with a
different (higher) phase velocity c than the high frequency
components. The theory of surface waves and its interpretation
is quite complicated and beyond the scope of a superficial
review paper such as this. The theory has been reviewed in

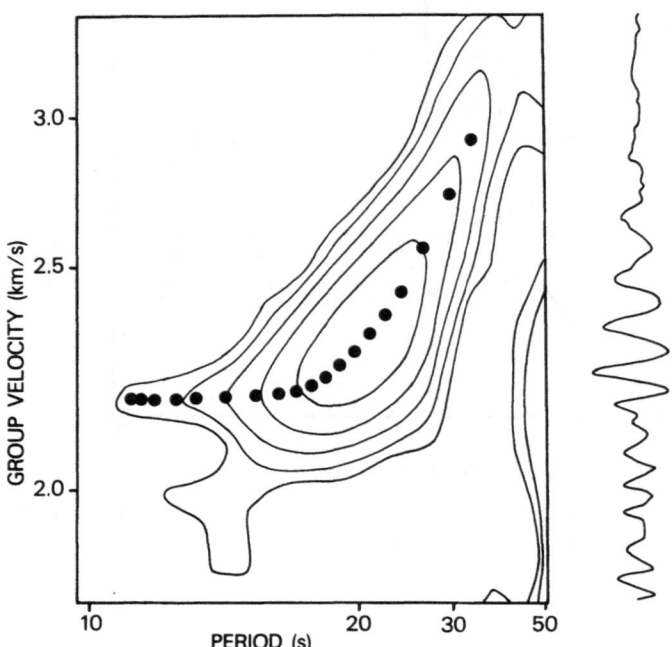

Figure 6   Multiple-filterdiagram for the recording in station
            BOL (Italy) of an event at the Greece-Albanian
            border – a distance of 893 km (after Nolet et al.,
            1978). Dots denote the measured values of the group
            velocities.

Aki and Richards (1980), Takeuchi and Saito (1972) and Nolet (1981),
the literature on dispersion observations by Knopoff (1972) and
Kovach (1978) and the reader is referred to these papers for
further detail. Here I shall limit myself to discussing the
usefulness of the method in the context of the investigation of
the crust and lithosphere.

    Tatham (1975) and Cloetingh et al. (1979) have studied the
use of surface waves for crustal investigations in sedimentary
basins and passive continental margins, respectively, and most
of their conclusions are valid for continental regions in general.

For the measurement of surface wave dispersion, use can be made of shallow earthquakes, preferably at short distance such that the event-station path travels in a geological province of one tectonic type only. The 15-s instruments of the WWSSN network are well-suited for the observation of the fundamental modes.

The easiest to measure is the group velocity, using the multiple-filter method, described in Dziewonski and Hales (1972). In this method, the digitized seismic signal is passed through a series of bandpass filters with increasing center frequencies, and the envelopes of the resulting signals are contoured in a time vs. frequency plot (figure 6). Interpretation of surface wave dispersion is not hampered by one or more low-velocity layers, and usually no identification problems occur, since the fundamental mode is a clearly recognizable wavetrain on long-period seismograms. Cloetingh et al. (1980) give a convenient set of standard curves for the interpretation of Rayleigh wave group velocities in the crust, where the inverse problem is quite non-linear. In general, group velocity measurements on WWSSN records may resolve the average Moho depth to a precision of 5 km or better, and may give useful estimates of the average S-velocities in the major layers of the crust.

Recently, methods have been developed for the measurement of higher modes of surface waves (Nolet, 1975). For this one needs an array of stations and digital registration is almost indispensable, which make higher modes rarely available for local, crustal studies. However, the resolution to be gained by the observation of higher modes is considerable, and the growing use of portable digital seismographs may make local higher mode analysis possible in the near future. Cara et al. (1981) have shown that this is possible in principle, using modes with periods around 2.5 seconds and local arrays in the western U.S.

CONCLUSIONS

Most of the methods that have been passed in review in this paper require just one seismological observatory in the neighbourhood of a seismic survey in order to obtain one or more important constraints on the Earth models. These constraints are usually in the form of travel times in the crust, but spectral ratio's and surface waves may even give a direct estimate of the Moho depth. The measurements are usually not difficult, nor time-consuming when compared to a refraction or reflection survey. In my view, seismic survey interpretations should be supplemented with global seismic data analysis whenever possible, in order to avoid errors due to misidentification of phases.
The 3-D method and the higher modes require arrays of seismic stations. In view of the recent developments in instrumentation,

the time may not be far away that these methods too will find
general application without excessive costs involved.

REFERENCES

Aki, K., A. Christofferson and E.S.Husebeye, 1976, Three-dimensional
    seismic structure of the lithosphere under Montana LASA,
    Bull.Seism.Soc.Am., 66, 501-524.
Aki, K., A. Christofferson and E.S.Husebeye, 1977, Determination of
    three-dimensional seismic structure of the lithosphere,
    J.Geophys.Res., 82, 277-296.
Aki, K. and H.K. Lee, 1976, Determination of three-dimensional
    velocity anomalies under a seismic array using first P arrival
    times from local earthquakes. 1. A homogeneous initial model,
    J.Geophys.Res., 81, 4381-4399.
Aki, K. and P.G.Richards, 1980, Quantitative Seismology, theory and
    methods, W.H.Freeman and Co., San Francisco, I and II, p.932.
Ben Menahem, A. and S.J.Singh, 1981, Seismic Waves and Sources,
    Springer Verlag, Berlin, p. 1108.
H.Bungum, S.E.Pirhonen and E.S.Husebeye, 1980, Crustal thickness in
    Fennoscandia, Geophys.J.R.astr.Soc., 63, 759-774.
Cara, M., J.B. Minster and R. Le Bras, 1981, Multi-mode analysis of
    Rayleigh-type Lg. Part 2: Application to Southern California
    and the northwestern Sierra Nevada, Bull.Seism.Soc.Am., 71,
    985-1002.
Cloetingh, S.A.P.L., G.Nolet and M.J.R.Wortel, 1979, On the use of
    Rayleigh wave group velocities for the analysis of continental
    margins, Tectonophysics, 59, 335-346.
Cloetingh, S.A.P.L., G. Nolet and M.J.R.Wortel, 1980, Standard
    graphs and tables for the interpretation of Rayleigh wave group
    velocities in crustal structures, Proc.Roy.Neth.Ac.Sci., B, 83,
    (1), 101-118.
Dziewonski, A.M. and A.L.Hales, 1972, Numerical analysis of
    dispersed seismic waves, in: Methods in Computational Physics,
    B.A.Bolt (ed.), Ac.Press, London, 11, 39-85.
Dziewonski, A.M. and B.H.Hager and R.J.O'Connell, 1977, Large-scale
    heterogeneities in the lower mantle, J.Geophys.Res., 82,
    239-255.
Ellsworth, W.L. and R.Y.Koyanagi, 1977, Three-dimensional crust and
    upper mantle structure of the Kilanea volcano, Hawaii,
    J.Geophys.Res., 82, 5379-5394.
Fairhead, J.D. and C.V.Reeves, 1977, Teleseismic delay times,
    Bouguer anomalies and inferred thickness of the African litho-
    sphere, Earth Plan.Sc.Lett., 36, 63-76.
Franklin J.N., 1968, Matrix Theory, Prentice Hall Inc., pp.292.
Frohlich, C. and M.Barazangi, 1980, A regional study of mantle
    velocity variations beneath E.Australia and the S.W.Pacific
    using short period recordings of P, S, PcP, ScP and ScS waves
    produced by Tongan deep earthquakes, Phys.Earth Plan.Int., 21,
    1-14.

Hirahara, K., 1977, A large-scale three dimensional seismic
    structure under the Japan islands and the Sea of Japan,
    J.Phys.Earth, 25, 393-417.
Jeffreys, H. and K.E. Bullen, 1940, Seismological Tables,
    Brit.Ass.Adv.Sci., London.
Jordan, T.H. and L.N. Frazer, 1975, Crustal and upper mantle
    structure from Sp phases, J.Geophys.Res., 80, 1504-1518.
Kanasewich, E.R., T. Alpaslan and F. Hron, 1973, The importance
    of S-wave precursors in shear-wave studies, Bull.Seism.Soc.
    Am., 63, 2167-2176.
Kennett, B.L.N., 1982, Seismic waves in stratified media, Cambridge
    University Press.
Knopoff, L., 1972, Observation and inversion of surface wave
    dispersion, Tectonophysics, 13, 497-519.
Kovach, R.L., 1978, Seismic surface waves and crustal and upper
    mantle structure, Rev.Geophys.Space Phys., 16, 1-13.
Kurita, T., 1973, A procedure for elucidating fine structure of
    the crust and upper mantle from seismological data,
    Bull.Seism.Soc.Am., 63, 189-209.
Leblanc, G.S.J., 1967, Truncated crustal transfer functions and
    fine crustal structure determination, Bull.Seism.Soc.A., 57,
    719-733.
Menke, W.H., 1977, Lateral inhomogeneities in P velocity under
    the Tarbela array of the Lesser Himalayas of Pakistan,
    Bull.seism.Soc.Am., 67, 725-734.
Mitchel, B.J., C.C. Cheng and W. Stauder, 1977, A three-
    dimensional velocity model of the lithosphere beneath the
    New Madrid seismic zone, Bull.Seism.Soc.Am., 62, 1061-1074.
Nolet, G., 1975, Higher Rayleigh modes in Western Europe, Geophys.
    Res.Lett., 2, 60-62.
Nolet, G., 1981, Linearized inversion of (teleseismic) data,
    in: The solution of the inverse problem in geophysical inter-
    pretation, R.Cassinis (ed.), Plenum Press, N.Y., 9-38.
Nolet, G., G.F. Panza and R. Wortel, 1978, An averaged model for
    the Adriatic subplate, Pure appl.geophys., 116, 1284-1298.
Nolet, G. and St.Mueller, 1982, A model for the deep structure
    of the East African Rift system from simultaneous inversion
    of teleseismic data, Tectonophysics, 84, 151-178.
Okal, E.A. and D.L. Anderson, 1975, A study of lateral inhomo-
    geneities in the upper mantle by multiple ScS travel-time
    residuals, Geophys.Res.Lett., 2, 313-316.
Phinney, R.A., 1964, Structure of the earth's crust from spectral
    behaviour of long-period body waves, J.Geophys.Res., 69,
    2997-3017.
Pilant, W.L., 1979, Elastic waves in the Earth, Elsevier,
    Amsterdam, pp.493.
Poupinet, G., 1979, On the relation between P-wave travel time
    residuals and the age of continental plates, Earth Plan.
    Sc. Lett., 43, 149-161.

Reasenberg, P., W. Ellsworth and A. Walter, 1980, Teleseismic
    evidence for a low-velocity body under the Coso Geothermal
    Area, J.Geophys.Res., 85, 2471-2483
Romanowicz, B.A., 1979, Seismic structure of the upper mantle
    beneath the United States by three dimensional inversion
    of body wave arrival times, Geophys.J.R.astr.Soc., 57,
    479-506.
Romanowicz, B.A., 1980, A study of large-scale lateral
    variations of P velocity in the upper mantle beneath
    western Europe, Geophys.J.R.astr.Soc., 63, 217-232
Sacks, I.S., J.A. Snoke and E.S. Husebeye, 1979, Lithospheric
    thickness beneath the Baltic shield, Tectonophysics, 56,
    101-110.
Sipkin, S.A. and T.H. Jordan, 1976, Lateral heterogeneity of
    the upper mantle determined from the travel time of multiple
    ScS, J.Geophys.Res., 81, 6307-6320.
Smith, W.T., 1970, S to P conversion as an aid to crustal studies,
    Geophys.J.Roy.astr.Soc., 19, 513-519.
Takeuchi, H. and M. Saito, 1972, Seismic Surface Waves, *in*:
    Methods in Computational Physics, B.A. Bolt (ed.),
    Ac.Press, London, pp.217
Tatham, R.H., 1975, Surface wave dispersion applied to the
    detection of sedimentary basins, Geophyics, 40, 40-55.
Vinnik, L.P., 1977, Detection of waves converted from P to SV
    in the mantle, Phys.Earth planet.Int., 15, 39-45.
Whitcombe, D.N., 1982, Three-dimensional seismic ray tracing for
    the forward modelling and direct inversion of teleseismic
    delay times, Geophys.J.R.astr.Soc., 69, 635-648.
Young, G.B. and L.W. Braile, 1976, A computer program for the
    application of Zöppritz's amplitude equations and Knott's
    energy equations, Bull.Seism.Soc.Am., 66, 1881-1885.

THE RELATION BETWEEN SUPERFICIAL AND DEEP STRUCTURE

OF THE CAUCASUS REGION AND POSSIBLE GEODYNAMIC MODEL

Victor N. Sholpo

Institute of Physics of the Earth
Academy of Sciences of the USSR
Moscow D-242

ABSTRACT

Longitudinal directions are clearly manifested in the present day structure of the Caucasus region, which formed during the Alpine geotectonic cycle (Mz-Kz). This might indicate the existence of a collision zone. However the presence of the transversal directions, which can be recognised both in the superficial and in the deep crust-upper mantle structure, are not less important.

Because the transversal direction occurs in the whole geological history and more strongly from the last orogenic (N-Q) stange, the activity of the deep crust cannot be explained as a simple collision between plates along a longitudinal line. Magmatic phenomena show that all geological history can be divided into two stages: the first one - geosinclynal stage with longitudinal activity of the deep zones of the crust; the second - orogenic stage with transversal activity. For both the stages of development the most probable mechanism for the evolution of the earth's crust are diapir-like movements on different levels in the crust and the upper mantle.

INTRODUCTION

Caucasus region occupies some isthmus between the Black Sea and the Caspian Sea basins, on the northern Edge of the eastern part of the Mediterranean Alpine mobile belt. The longitudinal zonation is very clear in this part of the mobile belt, and it appears in good agreement with the general trend of the whole belt.

The Caucasus is longitudinally divided into four major histo-

23

rical-structural units (from north to south): (i) the Precaucasian foredeep, (ii) the Greater Caucasus meganticlinorium, (iii) the Cura-Rion intermountain depression, (iv) the uplift of the Lesser Caucasus, which is the part of the internal Anatolian uplifted area of the Mediterranean belt. Such longitudinal division of the Caucasian region give us the possibility to think about the existence here of an ancient collision zone. This collision zone may be considered as a responsible structure for the present-day observed structure of this region. Many investigators advanced such supposition (Adamia et al., 1977, Adamia et al., 1980, Dewey et al., 1973, Gamkralidze, 1976).

But the presence of the significant transversal structures, the most important of them being the "Transcaucasian uplift" which crosses the whole Mediterranean belt, makes such a supposition very problematic. For the purpose to reach some affirmative decision about the problem of the evolution of the earth's crust in this region it is necessary to consider all the conceivable actual data about present-day geological structure, viz: the distribution of the folding and other deformations of the rocks, the main stages of the geological development, the evolution of the magmatic phenomena during the Alpine cycle, the data on the deep structure and the physical conditions in the inner part of the crust and upper mantle.

The suitable geodynamic model it is possible to elaborate after compatible consideration all this data only. It was important to consider especially the structure and development of the Greater Caucasus because the processes of the geosinclynal development occured here in a more conclusive form as the complete meganticlinorium was formed.

GEOLOGICAL STRUCTURE OF THE REGION

As it was mentioned above, the Caucasus can be divided from a structural and orographical point of view into four major units (fig. 1) which extend in longitudinal direction. Each of this units is divided by transversal faults or deep flexures into several blocks of different dimensions, thus an almost orthogonal net of lineaments makes the Caucasus a complicated pattern of blocks of different orders.

Precaucasian foredeep confines the Greater Caucasus meganticlinorium in the north and east, and separates it from the Scythian epi-Hercynian platform. Along the trend this foredeep is divided into three parts, which are different in the inner structure and in the level of prealpine basement surface. Stavropol uplift divides the foredeep into two unequal parts: the western one - the Kuban basin and the eastern one of the - Terek-Caspian trough. The prealpine basement in the Stavropol uplift is I.5 - 2.0 Km deep under the cover of Mesozoic and Cenozoic sedimentary rocks. The level of

Fig. 1. Block-structure of the Caucasus region.
1 - 6-Alpine folded area: 1 - prealpine uplifted basement;
2 - relative uplifted step- blocks; 3 - relative non uplifted step-blocks; 4 - fore and intermountain deeps;
5 - young volcanic flow; 6 - volcanic edifice, laccolite;
7 - transverse flexure of I order; 8 - transverse flexure
of II order; 9 - border line of main uplifts and basins
inside folded area; I0 - border line of structural blocks;
11 - epi-Hercynian platform; I2 - uplifted part of epi-
Hercynian platform.

the prealpine basement in the Kuban basin changes in depth from
3-4 Km on the east to 7-8 km on the west. This change occurs along
the steep transversal flexure or deep fault zone. The gentle basin
of Kuban foredeep has an asymmetrical structure: the southern limb
is more steep than the northern one: in the south the intensive
uplifted mountain chain of the Greater Caucasus appears. For this
reason the axial part of foredeep was shifted towards the south.

The Terek-Caspian trough is wider than the Kuban basin and the
subsidence is larger. The latter reaches its maximum value (IO Km)
at the estuary of the Sulak river. In the western part of this
trough (Terek basin) the sedimentary cover is distorted by two longitudinal elongated anticlinal folds, where diapiric phenomena occurred. Also along the Caspian shore two fault lines are known along
which several dome-like young anticlines are situated.

The block structure of the Greater Caucasus meganticlinorium
is one of its major distictive features. The blocks of the Greater
Caucasus differ greatly as far as the nature of the rocks, the
degree of metamorphism and the deformations of sedimentary units.
Despite the block heterogeneity, however, the general anticlinorium
pattern may be observed all over the Greater Caucasus. Earth's crust
blocks of different origin and composition are integrated in the
Greater Caucasus meganticlinorium giving origin to a complete and
regular structure that formed during the Alpine geotectonic cycle.
The eastern part of meganticlinorium is relatively subsided and the
central axial zone is occupied by blcoks composed of the lower part
of the Alpine sedimentary complex (Lower and Middle Jurassic terri-
genous sediments). To the south and north from the axial zone the
surface of the neighbouring blocks is composed of rocks of the middle part of the Alpine sedimentary complex (upper Jurassic-Cretace-
ous), mostly carbonatic rocks (flysch zone on the southern slope,
Daghestan Limestone on the northern one).

In the western part of the meganticlinorium the axial zone is
occupied by a block with an exposed Prealpine basement (fig. 2,1-1).
Rocks of the lower part of the Alpine complex prevail in the blocks
north and south of this central zone (Labino-Malkinskaya zone in
the north, Svanetija - in the south). While these rocks units are
only gently deformed in the north, to form a slightly titled monocline, in the southern zone they form a highly deformed anticlinorium. Here the units assigned to the middle part of the Alpine complex exclusively compose the meganticlinorium limbs and become more
abundant in the adjacent zones beyond the Greater Caucasus, in the
so called Zacaucasian Median Massif and in the Scythian platform.
The axial zone of the Alpine structure is therefore displaced
southward in the western part of the Greater Caucasus as compared
with the eastern part, and does not coincide here with the main
axis of meganticlinorium. This situation greatly emphasizes its

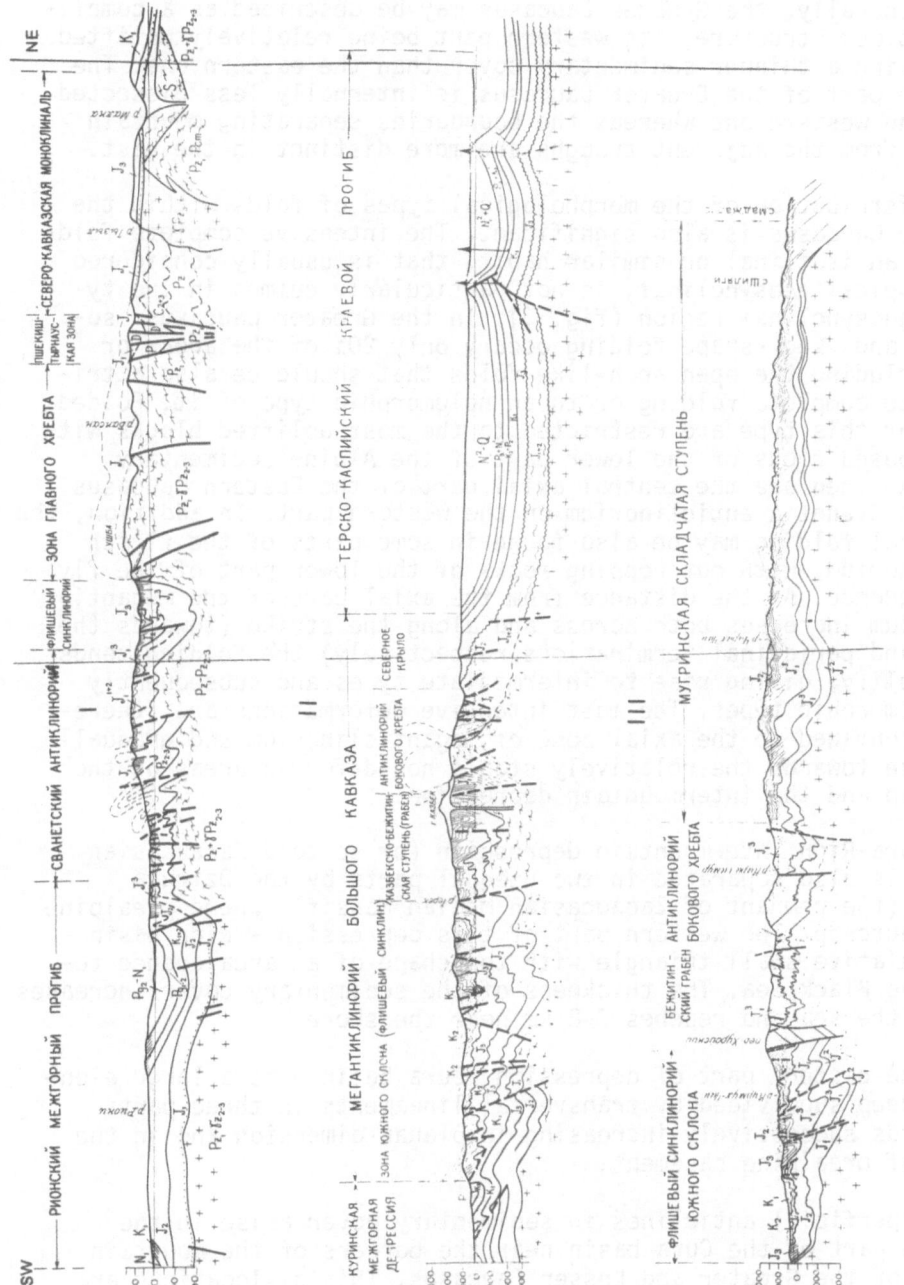

Figure 2

27

general asymmetrical pattern, i.e. the wide and gentle northern limb and the narrow and steep southern one.

Generally, the Greater Caucasus may be described as a complicated block structure, its western part being relatively uplifted and having a thinner sedimentary cover than the eastern one. The eastern part of the Greater Caucasus is internally less dissected than the western one whereas the boundaries separating mountain chains from the adjacent troughs are more distinct in the east.

Distribution of the morphological types of folds within the Greater Caucasus is also significant. The intensive complete folding of an isoclinal or similar habit, that is usually considered as a typical geosynclinal, is not particularly common in the typical geosynclinal region (fig. 3). In the Greater Caucasus isoclinal and keel-shape folding occury only 20% of the area, or 38% including the open arch-like folds that should be also attributed to complete folding or to an holomorphic type of it. Folded areas of this type are restricted to the most uplifted blocks with the exposed rocks of the lower part of the Alpine sedimentary complex. Then are the central axial part of the Eastern Caucasus and the Svanetia anticlinorium on the western part. In addition, the isoclinal folding may be also found in some parts of the flysch synclinorium, with outcropping rocks of the lower part of the flysch sequence. As the distance from the axial part of the meganticlinorium increases both across and along the strike (towards the limbs and periclinal terminations respectively) the folding tends to simplify, giving rise to intermediate types and subsequently to idiomorphic types. The most intensive deformations are, therefore, confined to the axial zone of meganticlinorium and gradually decrease towards the relatively stable non-deformed areas of the foredeep and the intermountain depression.

Cura-Rion intermountain depression (or so cold Zacaucasian basin) is also separated in two unequal parts by the Dzirula Massif (the remnant of Zacaucasian Median Massif), where prealpine rocks outcrop. The western part of this depression - Rion basin - is a relative small triangle with the shape of an area opened toward the Black Sea. The thickness of the sedimentary cover increases toward the sea and reaches 7-8 km near the shore.

The eastern part of depression -Cura basin - is a large elongated deep subdivided by transversal lineaments in three parts eastwards successively increasing in planar dimension and in the depth of prealpine basement.

Superficial anticlines in sedimentary cover arise in the eastern part of the Cura basin near the borders of the mountain chains of the Greater and Lesser Caucasus. This dislocations are

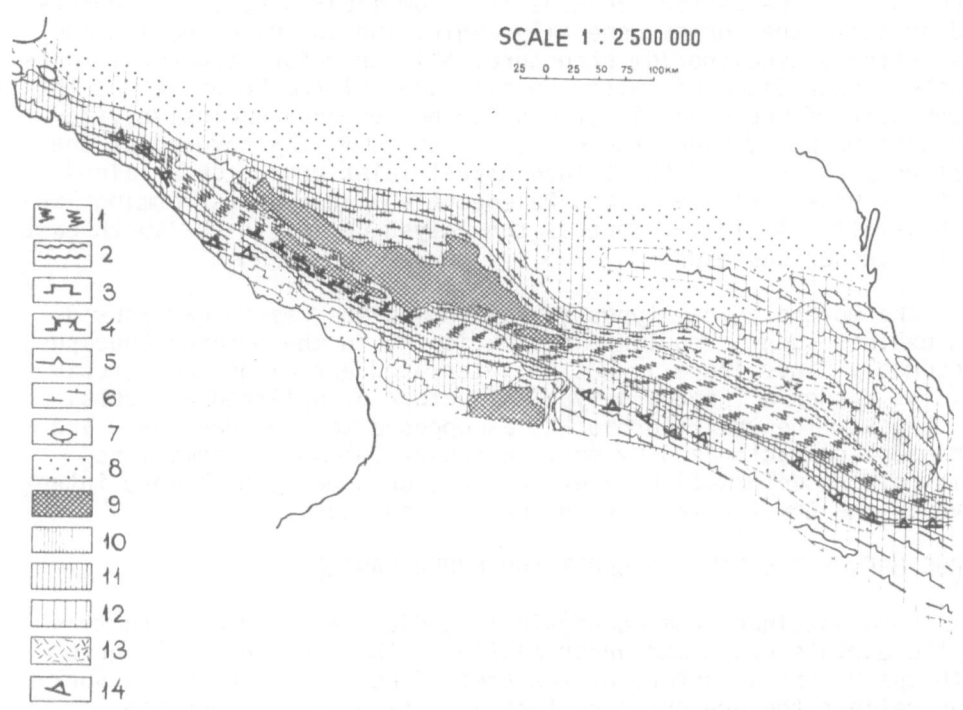

SCALE 1 : 2 500 000

25 0 25 50 75 100KM

Fig. 3 - Distribution of type of folding in the Great Caucasus.

1 - 7 - type of folding: 1 - isoclinal and keel-like;
2 - open arched; 3 - box-fold; 4 - box-fold with strong
disharmonic phenomena; 5 - ridge-like; 6 - monocline;
7 - dome-like, brachy-anticline; 8 - sedimentary cover
almost without deformations; 9 - outcropping prealpine
basement; I0 - Lower-Middle Jurassic sedimentary units;
11 - Upper Jurassic-Paleogene sedimentary units; I2 -
Neogene-Quaternary sediments; I3 - young (N-Q) volcanic
flows; I4 - overthrusts and frontal zones of nappes.

connected with deep fault zones in the basement.

In the Lesser Caucasus, on the contrary to the Greater Caucasus, there are no single axial zone and marginal parts to form an anticlinorium structure. In the uplift of the Lesser Caucasus a number of longitudinal anticlinal and synclynal zones may be found, usually separated by large regional faults that are, as a rule, deep sutures developed over a long period of time. The general structure of the Lesser Caucasus is a complicated system of horsts and grabens, the former generally exhibiting an anticlinorium and the latter a synclinorium structure. All the major structural zones of the Lesser Caucasus, with the exposures of the Mesozoic and the lower part of the Cenozoic strata may be observed exclusively in the eastern part, since the western area is mostly covered of the western part there is the folded Ajar-Trialet region underlained by thick Upper Cretaceous and Paleogene volcanic units forming large box-folds. A similar zone is the Talysh uplift in the south-east of the Lesser Caucasus.

The morphological types of folding in the Lesser Caucasus do not exhibit such a regular distribution as in the Greater Caucasus, presumably since there is not an integrated anticlinorium structure, but rather a "keyboard-type" structure of uplifted and subsided blocks. The uplifted blocks, as opposed to subsided ones, are generally characterized by more intensive folding deformations. This folding deformations are, however, presumably much more intensive in the Greater than in the Lesser Caucasus.

STRUCTURE OF THE EARTH'S CRUST AND UPPER MANTLE

Data obtained from geophysical studies on the deep structure of the earth's crust and upper mantle of the Caucasus well agree with the recent structure of the area. Since the geophysical surveys reflect the present structure and state of deep horizons, they are better correlated with the latest (orogenic) stage of development. The structure of the crust had been created, however, throughout its whole evolution; thus the earlier stages of its formation can be associated with certain pecularities of the deep structure.

As one can see from any map of the earth's crust thickness, its (fig. 4) changes of generally correspond to the major structural elements of the region. It can be noted that also the regular changes in the crust thickness take place inside the major structural units. The map shows that under the meganticlinorium of the Greater Caucasus the deep layers also break down into some ovals or lenses corresponding to those blocks that are observed in the surface structure. The greatest thickness of the earth's crust is typical for Central and Eastern Caucasus blocks, but they are

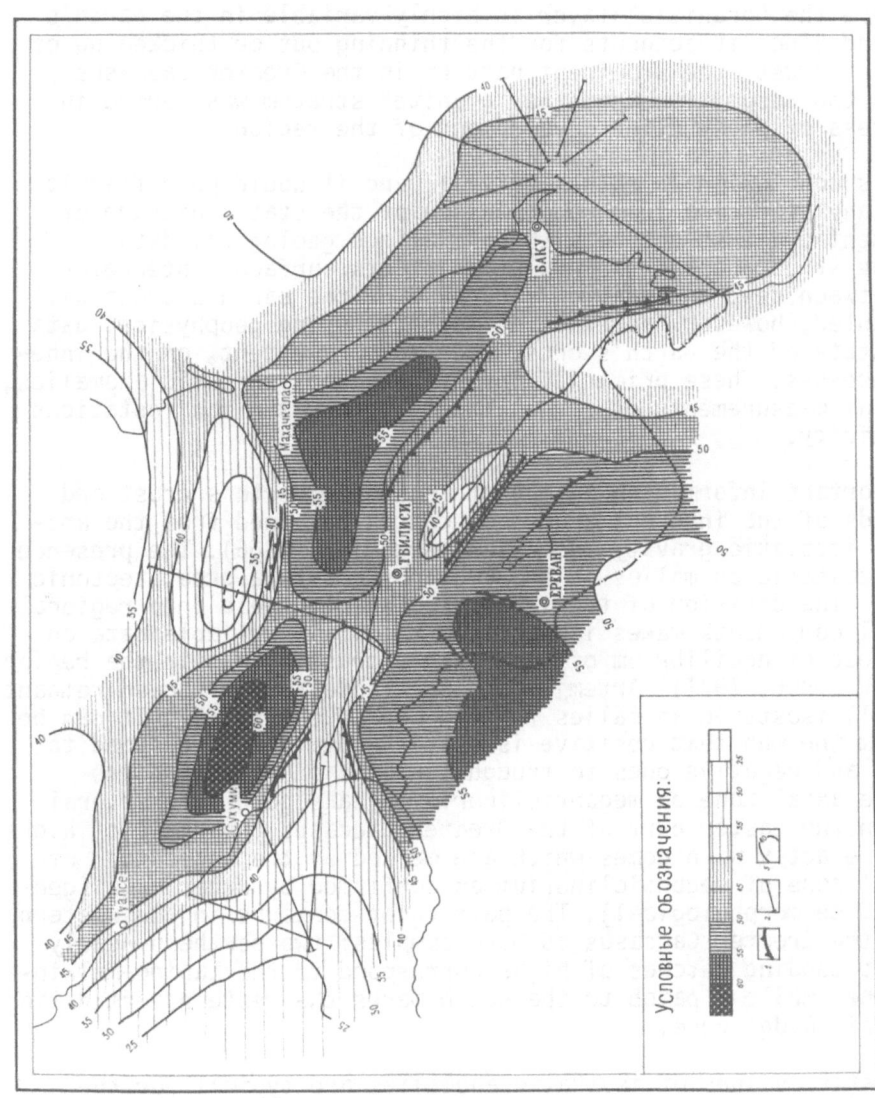

Fig. 4   Scheme of the thickness of the crust. Scale of thickness with steps of
5 Km 1 - fault zone; 2 - DSS profiles; 3 - isolines of thickness.

divided by a neck that coincides with a transverse flexure separating the blocks. The relative decrease in the crust thickness is recorded in the periclines of the meganticlinorium, between the blocks of the North-Western and Sout-Eastern Caucasus.

Since the "granitic" layer is highly variable in the earth's crust, and since it accounts for the thinning out or thickening of the whole crust, the prominent pattern in the Greater Caucasus supports the assumption that the "granite" stratum was formed in the process of geosyncline development of the region.

This conclusion is rather trivial, and it would be difficult to find anything else from a comparison of the static picture of the present crust structure with generalized geological data about the structure and the evolution of its surface. Interrelations between deep and shallow structure of the earth's crust will be presented, however, somewhat differently using geophysical data on the state of the earth's crust and, to some extent, on the inherent processes. These primarily involve data on isostatic anomalies, as well as measurements of recent thermal flows and manifestations of seismicity.

Important information on the state of the earth's crust and the trends of the inherent processes can be obtained from the analysis of isostatic gravity anomalies (Artemjev, 1966). The presence of the isostatic anomalies in the region suggests a great tectonic activity. The division of the isostatic anomaly field into regional and local components makes it possible to obtain reliable data on the degreee of equilibrium of the earth's crust blocks in the region (Artemjev, 1966, 1971). Artemjev and Balavadze (1973) showed remnant (or local) isostatic anomalies for the Caucasus (fig. 5). It can be seen from the map that positive isostatic anomalies correspond to uplifts, and negative ones to troughs. A belt of gravity highs along the axial zone of meganticlinorium breaks down into several ovals. In the easter part of the Greater Caucasus the ovals of highs coincide exactly with domes which are marked by the undulation of the axial zone of meganticlinorium as confirmed by other data (geochemical, geomorphological). The pattern of contours in the western part of the Greater Caucasus is more comples: two northern nearly east-west tending patches of highs correspond to the Labino-Malkinskaya zone, while a patch to the south marks the western termination of the Main Ridge zone.

Negative values of isostatic anomalies are typical for the periclinal parts of the meganticlinorium. Thus, the map of local anomalies is closely associated with the Alpine structure of the Greater Caucasus, and, in fact, it is a generalized structural scheme. Consequently, if the gravity equilibrium was disturbed and is still being disturbed ny neotectonic movements, the data indica-

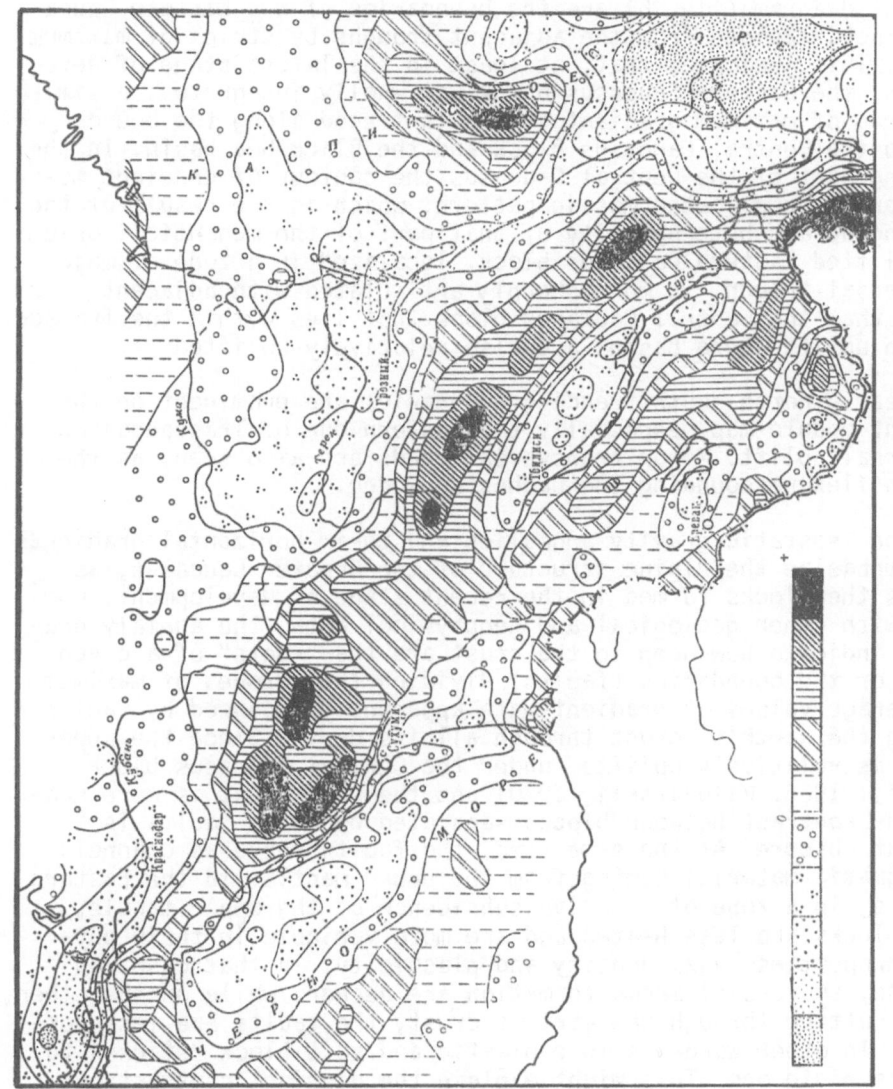

Fig. 5 – Scheme of the remaining isostatic anomalies (after M.E. Artemjev and B.K. Balavadze, 1973)

33

te that the crust block movements were controlled by the structure formed at previous stages.

The horizontal gradients of remnant isostatic anomalies mark the boundaries between blocks of the earth's crust with a different gravity equilibration and ensure a quantitative assessment of the contrast of the boundaries. Particularly distinct on the horizontal gradient diagram (fig. 6) are the boundaries of the Eastern Caucasian block, divided from the adjacent troughs by strips of maximum values. This indicates that, at least in the latest stage of development, the Eastern Caucasus block was wholly integrated. A similar strip of maximum gradient values is traced along the boundary between the Central Caucasus block and the Black Sea basin. In the remaning part of the Central Caucasus the contour of gradient module form a complicated mosaic pattern, which is the result of the breaking up and heterogeneity of that part of the meganticlinorium. The uplifted Eastern Caucasus block, localized in a zone of major transversal trough, is more sharply articulated with adjacent basins than the Central Caucasus block that lies within the Transcaucasian uplift where basins are also relatively uplifted.

Nearly north-south trending boundaries are prominent on the gradient module map, especially the eastern one of Trascaucasian trasversal uplift. Minor transversal bends are also seen, as the eastern flexure bounding the Daghestan wedge.

The isostatic gravity anomalies and their horizontal gradients thus emphasize the Alpine structure of the Greater Caucasus, as well as the blocks formed in the recent stage of development. Compared with other geological and geophysical data, the anomaly gradients indicate how deep in the crust are the "roots" of a given block or the boundaries (faults) dividing them. Zones of maximum and average values of gradients are apparently produced by faults cutting the earth's crust through all its depth. Since the upper mantle is relatively uplifted under the Central Caucasus block (Sorskij, 1966; Malovitskij, 1970) and the here crust is more heated, the contrast between blocks separated by large faults is somewhat obscure. At the same time, the faults serve as channels for magmatic material coming from the upper mantle. In the Eastern Caucasus, in a zone of relative subsidence of the upper mantle, crust blocks are less heated and are more variable in their physical properties, viz. density and plasticity. In that case, on the hand, sources of magma formation are deeper, while on the other, though cutting through the earth's crust, the faults are less permeable. In other words, a more plastic and soft block "sticks" to the more rigid one. This might explain the distinct contrast between the conjugation of the Eastern Caucasus and adjacent regions, where actually no young volcanism is involved.

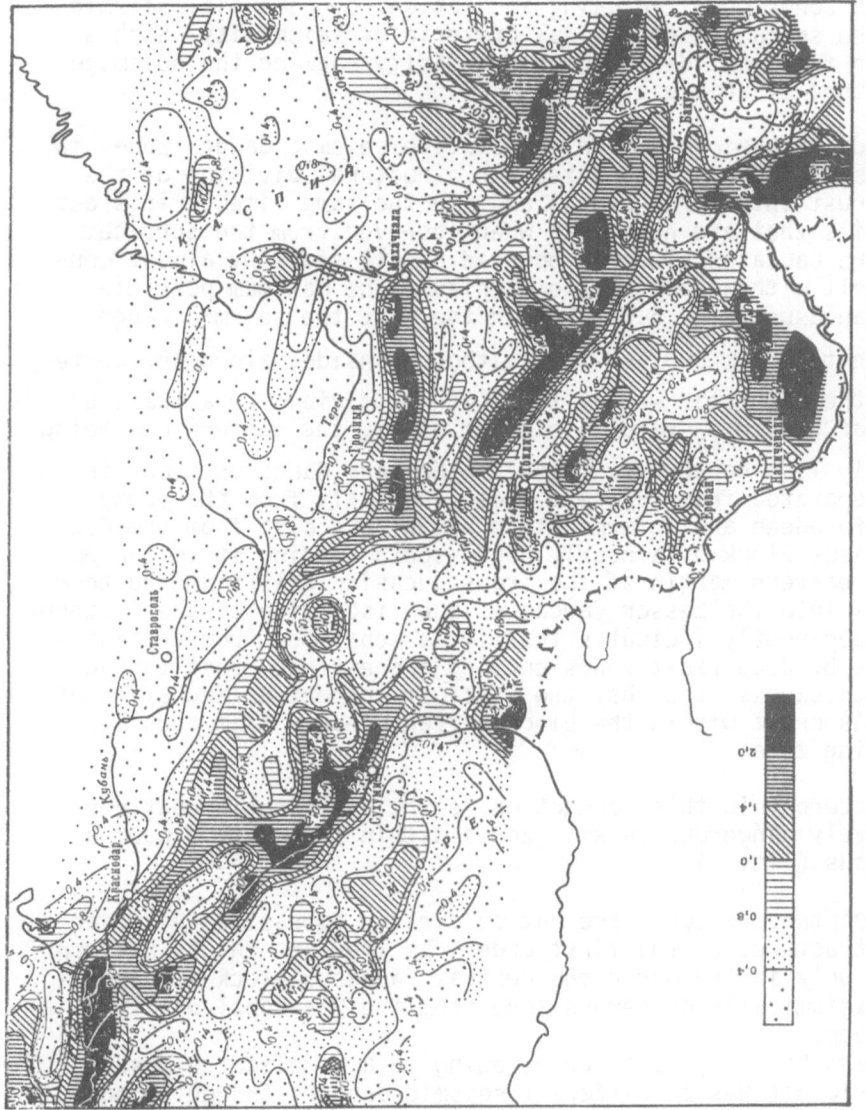

Fig. 6 – Scheme of horizontal gradients of the remaining isostatic
anomalies (after M.E. Artemjev and B.K. Balavadze, 1973)

35

Since seismic activity is a manifestation of endogenous tectonic processes, all the informations about the intensity of earthquakes in the area are directly related to the earth's crust structure and the character of the inherent processes. It seems reasonable to obtain information on the earth's crust structure not from a map of recorded epicentres, but rather from a prognostic map of earthquake zones compiled using indications of potential seismic danger, the so-called geological criteria of seismicity. Such a map is, in fact, an ideal map of seismicity showing the seismogenic structure of an area.

The distribution pattern of relative seismic danger zones in the Caucasus (fig. 7) emphasizes the structural division of the earth's crust into large blocks and, in the same time, brings out some details that are not clear when observed from the surface. The Eastern Caucasus block is bounded on all sides by almost continuous belt with an earthquake magnitude the maximum possible for the Caucasus (M_{max} 6.5). Within the belt the seismic danger drops down to M_{max} 5.0 and lower, though inside, along the western trend of the Daghestan wedge, the block is divided by a transversal zone of high M_{max} values into two parts, the eastern one being seismically less dangerous. Thus, the Eastern Caucasus block is sharply separated from adjacent areas, not only from the young troughs (foredeep and intermountain basin) but also from the Central Caucasus block by a nearly north-south tending strip traced along the eastern margin of the Transcaucasian uplift and further southwards into the Lesser Caucasus. Such isolation of the Eastern Caucasus apparently indicates that it is separated from the adjacent areas by deep fault zones cutting the earth's crust through all its thickness, and that the physical properties and state of the earth's crust within the block differ greatly from those in neighbouring zones.

Of interest in this connection is the distribution pattern of relatively dangerous seismic zones within the western part of the Caucasus (fig. 7).

The earthquake zones are not so clearly emphasized by major surface structures of the first order, as they are in the Eastern Caucasus. Only in the north the Central Caucasus block is separated by a seismically dangerous zone from the Stavropol uplift and foredeep.
Its southern boundary is obscure, owing to a complicated mosaic of isometric patches of different seismic danger, though among them are high potential seismic zones.
Thus, the western part of the Caucasus is more heterogeneous in its internal part than the Eastern Caucasus, and the Greater Caucasus seems not to be so sharply isolated from neighbouring zones as in the east.

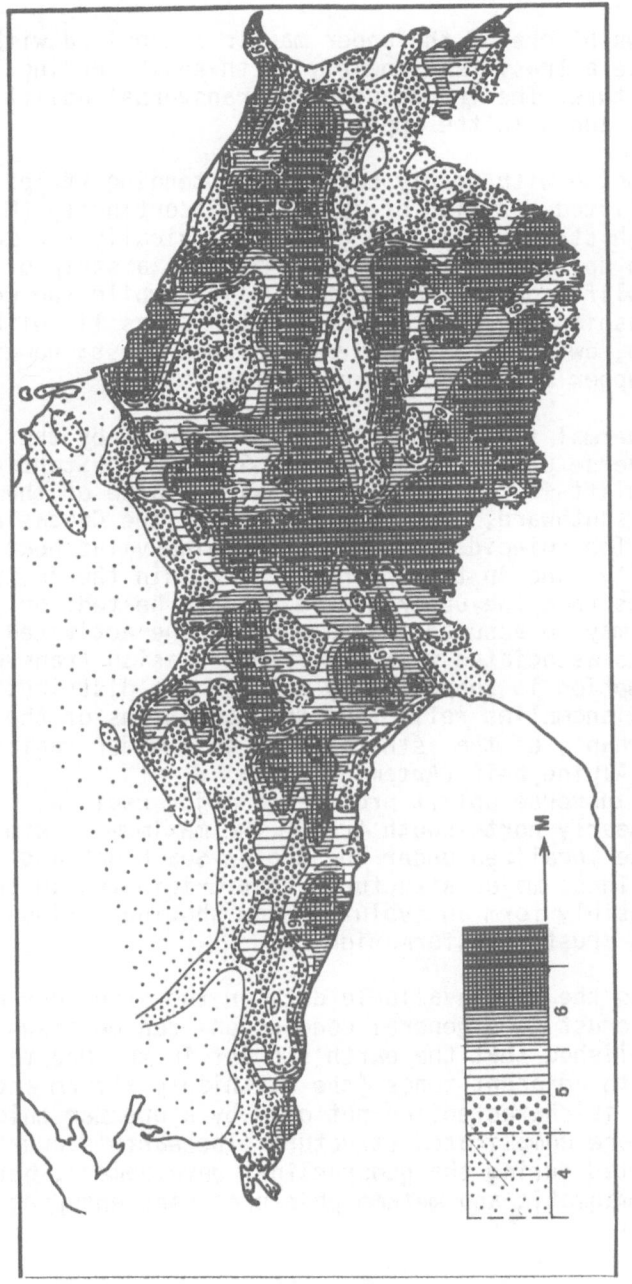

Fig. 7 – Scheme of prediction of the seismoorigin zones in magnitudes (compiled by B.A. Borissoff and G.I. Reisner)

On the basis of records of major earthquakes, using a special technique (Vinnik, Lenartovich, 1976), it became possible to identify horizontal unhomogeneities in the upper mantle.

The heterogeneous blocks of the upper mantle recognized within the Caucasus emphasize a trasversal, nearly north-south rending, zonation of the structure. The Transcaucasian transversal uplift produces low-velocity zones in the upper mantle.

These zones coincide with nearly north-south tending strips of the relatively uplifted position of the Moho discontinuity (Malovitskiy, 1970), though structurally and geomorphologically the zones may not coincide with uplifts. A low-velocity zone in a strip of the Transcaucasian uplift corresponds to the uplift, while the western coast of the Caspian Sea with adjacent water areas lie within a transverse downwarp, owing to a relatively thinned crust; however the boundary of the upper mantle is there uplifted.

Somewhat high thermal flow values are also produced by the Transcaucasian transverse uplift (Lubimova et al., 1973). Starting from the Stavropol uplift in the north, the maximum value of the thermal flow extends southward, cutting through all the Caucasian structures (fig. 8). The coincidence between the high-value heat flow and a low-velocity zone in the upper mantle within the Transcaucasian uplift suggests that the upper mantle is more heated, or even melted. Thus it may be assumed that a zone of the activated or "excited" mantle is associated with the Transcaucasian transverse uplift. This assumption is supported by data on the distribution of regional isostatic anomalies reflecting the conditions of the upper mantle and, perhaps; of the asthenosphere in a wider region of the Mediterranean Alpine belt (Artemjev, 1971).
The Transcaucasian transverse uplift produces a major regional maximum of the same nearly north-south trend. The maximum is similar in nature to those localized under the Aegian Sea block and the Pannonian basin. These major structures that originated in the neotectonic stage possibly form an evolution row showing various stages in the earth's crust transformation.

After considering the main available data on the structure of the Caucasus earth's crust some general conclusions can be drawn. It is definitely established that the earth's crust in the Greater Caucasus, compared with adjacent zones (the Scythian platform and Intermountain trough) is characterized not only by a greater thickness, but also by a more complicated structure. The earth's crust was not just accumulated during the geosynclinal development, but was also reworked by magmatic and metamorphic processes ensuring a

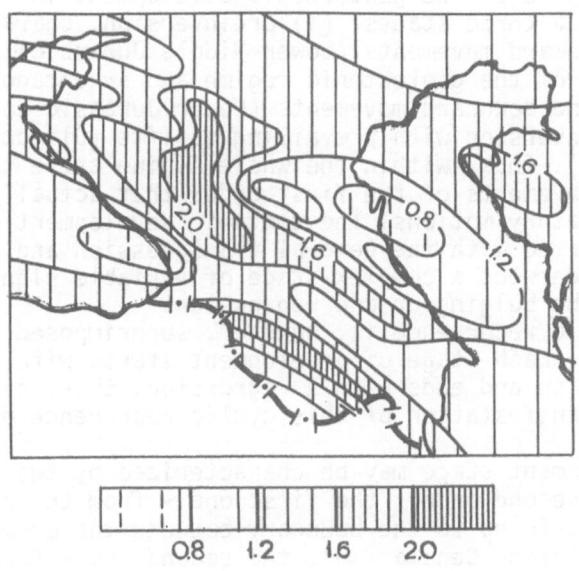

Fig. 8 - Scheme of the heat flow (after E.A. Lubimova and oth., 1973).
Isolines of the heat flow in 10^{-6} cal/cm^2.sec

thicker upper stratum, commonly called "granite". The matter of the rocks making up the earth's crust structure has been rather reworked and completely changed. In other words, it was mixed up during the intensive movements of the Alpine geosynclinal cycle. At the same time, seismic boundaries in the crust indicate changes in the physical properties of the matter and, perhaps, boundaries or extension of metamorphic processes.

DEVELOPMENT OF VERTICAL MOVEMENTS DURING THE ALPINE CYCLE

The Alpine cycle of the geotectonic development in the Caucasus may be divided into three stages: (i) preinversion, characterized by prevailing downward movements (Lower-Middle Jurassic); (ii) a partial inversion of the geotectonic regime and an antagonism between the upward and downward movements (Upper Jurassic-Eocene); (iii) a general inversion with prevailing uplifts (Oligocene-Anthrogene) (fig. 9). Thus, within the whole Alpine cycle there were clearly regular movements of the first order that actually formed a cycle of oscillatory motions. The tectonic development of the region, which started with the general transgression and related subsidence and underwent a complex stage of variable sign movements, is completed by the bulging up and regression.
This general cyclic recurrence is, however, superimposed by that of a higher order. Each stage of development starts with a major regional transgression and ends with a regression; thus, each individual stage is a manifestation of this cyclic recurrence of the second order.
The second development stage may be characterized by two oscillatory cycles of the second order: the first one - from the beginning of the Upper Jurassic up to the boundary between the Lower and Upper Cretaceous (Albian- Cenomanian), the second one - from the middle of the Upper Cretaceous to the end of the Eocene. This oscillatory cycles of the second order affected not so much the distribution of sedimentary thickness, but rather the composition of the sediments. Trasgressive sedimentary series originate at the beginning of each cycle. These may be presumably characterized by non--compensated downwarping (Middle Liassic argillaceous sediments, Upper Jurassic-Valanginian flysch formation).
Each cycle is completed by accumulation of the regressive sedimentary successions with clear indications of "overcompensation" - coal bearing sediments, lagoon facies.

Longitudinal zoning of the Caucasus may be clearly traced througout the Alpine cycle. The periphery of the region, i.e. the southern part of the Scythian platform and the Transcaucasian massif are subjected to oscillations of smaller amplitude compared with the internal part of geosynclinal zone of the Greater Caucasus. This longitudinal zoning is underlined by particular formations typical for each of the zones. The internal mobile zone is notable for geosynclinal formations (slate, flysch), whereas the

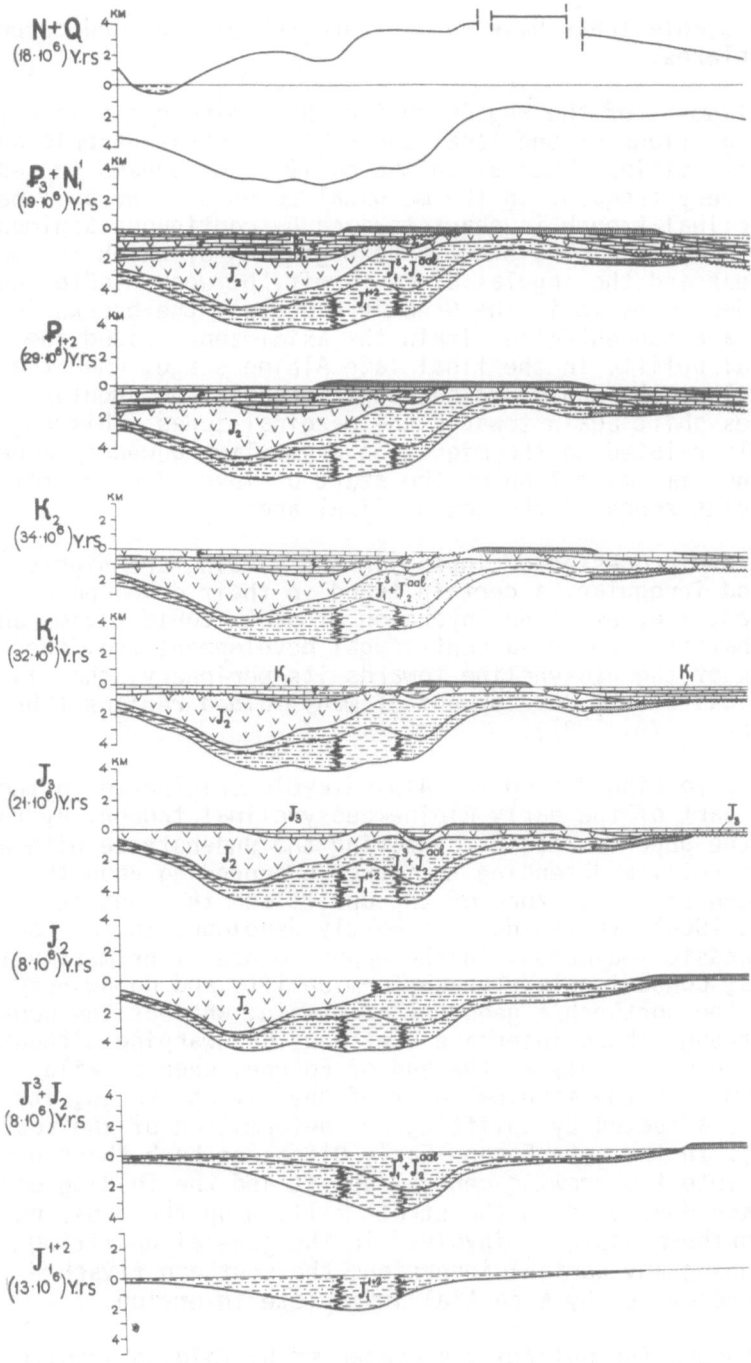

Fig. 9 - Structural- historical cross-sections through the Great Caucasus.

marginal more stable areas have subplatform terigeneous and carbonate rock complexes.

The development of the region during the Alpine cycle is governed by time migrations of the local and regional stratigraphic and angular disconformities. Whereas in the early stage breaks in sedimentation are very frequent in the marginal zones and the axil part of the geosynclinal trough is characterized by continuous sedimentation, in the middle Alpine stage, the beginning of which is, marked by the break and the angular disconformity between Middle and generally Upper Jurassic in the Greater Caucasus, the breaks in sedimentation are concentrated within the axial zone around the growing central uplift. In the final late Alpine stage, when the uplifting involves the whole region, stratigraphic and angular disconformities shift again towards the marginal zones. This regularity is closely related to the migration of maximum downwarp zones and thus to the time migration of the start of inversion and folding in different zones of the geosynclinal area.

Although the processes of inversion and folding are highly complicated and irregular, a certain trend in their development may be outlined, i.e. both the inversion of geotectonic regime and folding are characterized by a centrifugal development from the internal zones of the geosyncline towards its periphery, that is generally typical of the development of many folded regions (Khain, 1973, Beloussov, 1978, 1981).

The earler folding during the Alpine cycle originated in the central axial part of the early Alpine geosynclinal trough. By the beginning of the Upper Jurassic, overlying the understrate with an angular unconformity and tending to increase depending upon the distance between the axial zone of the uplift and the contact (Sholpo, 1962, 1964), the folds were mostly developed in the Lower and Middle Jurassic sediments. In the Upper Jurassic, primarily in the Cretaceous, consedimentary brachyform uplifts and downwarps are formed in the northern a geosynclinal basin, whereas the southern flysch trough shows intensive continous downwarping without any structural growth. Only at the end of Eocene, when a "wild flysch" is formed in the marginal part of the flysch trough, its central part is affected by uplifting and deformation of the rock (Leonov, 1975). In the Late Eocene-Early Oligocene both these basins involved into the growing central uplift and the folding of the second stage developed in the strate filling up these basins. Whereas the northern basin is involved in the general uplifting without undergoing any partial inversion, the southern flysch trough is characterized by a partial incomplete inversion.

In the Miocene the uplifting accompanied by folding involves certain parts of the foredeep and intermountain troughs. Pre-Upper Pliocene movements were responsible for brachyfolds in the so-called "Tertiary piedmonts", fringing the meganticlinorium of the Greater Caucasus.

The transversal zoning of the Greater Caucasus is also clear. At all stages of the Alpine cycle the Western Caucasus underwent a less intensive downwarping than the Eastern Caucasus. This transversal division of the Caucasus region is superimposed on the predominant longitudinal zoning, as appearing in "trasparence" through it. The transversal may become more intense (Toarcian-Aalenian, Neocomian and particularly Neogene-Quaternary time) or weaker at different stages of development. The present structure of the Caucasus is characterized by a renewal of intensive movements in the Trascaucasian uplift.

EVOLUTION OF ALPINE MAGMATISM

Although the magmatic rocks formed in the Greater Caucasus during the Alpine cycle are extremely various in composition, main features of the evolution of the magmatic phenomena may be obtained. The Alpine magmatic cycle in the Greater Caucasus is divided into two stages: the first one includes the Early, Middle and Late Jurassic, the second started at the boundary between the Early and Late Cretaceous and continued up to the Quaternary (Borsuk, 1974, 1977). Typical of the first stage was a common geosynclinal succession in the change of the magmatic formation within the mobile zone: a transition from basic rocks to acid one. The change in the composition of formations coincides exactly in time of the partial inversion of geotectonic regime. At the same time, following from one geotectonic zone to another across the strike of the geosynclinal area, there is a typical change of magmatic formations controlled by deeper sources of magma in the direction from the mobile zone to the platform.

The second stage is characterized by a more uniform petrology of the rocks of the magmatic formations associated with all structural-tectonic zones, irrespective of their prehistory. This fact, as well as the confinement of magmatism at that stage to the Transcaucasian uplift, suggests that magmatic activity at that stage was not directly related to the geotectonic evolution of the Greater Caucasus itself, but was rather controlled by a more general and presumably deeper process involving the whole Alpine mobile belt. This is also supported by the fact that magmatic formations of the similar composition manifested at that time not only in different zones of the Great Caucasus, but also in all the other tectonic zones of that segment of the Alpine belt. In addition, the petrology of the magmatic rocks of that stage, and primarily, their higher potassium alcalinity, compared with that of the rocks of previous stages, is indicating that magma generation levels became deeper at that time.

As it had been pointed out, however, the Alpine cycle of the Caucasus is divided in terms of tectonics into three stages: prein-version subsidence, partial inversion, and an orogenetic stage. The diagram (fig. I0) shows the correlation of stages of vertical tectonic movements in different structural-tectonical zones of the Greater Caucasus with principal stages of magmatic evolution. The diagram vertically presents the geochronological scale, showing the absolute duration of time segments. Horizontally are the struc-tural-tectonic zones of the Greater Caucasus corresponding to a cross-section through the Central Caucasus block (south to the left). The configuration and relative dimentions of subsidence and uplifted zones for different areas of the Greater Caucasus are gi-ven for each time segment on a tentative profile.Also the average rates of subsidence and the differentiation coefficient are shown. Manifestation of magmatism are designated by trianglles: a thick black line indicates acid formations, double thin lines - basic formations. Vertical arrows indicate the age limits of origin of formations, and the hatched segments inside triangles tentatively designates the amount of potassium alkalinity: the closer the seg-ment to the base of triangle the higher is the potassium alkalini-ty. Finally, the low line on the diagram indicates the age and de-gree of consolidation and "rigidity" of the Prealpine basement of each zone: the more dense is the vertical hatching in the line, the more consolidated in the basement.

The diagram shows clearly the regularities of magmatic evolu-tion discussed above. In the first stage the magmatism is variable and specific to each of the zones, and the age of magmatic forma-tions increases from south to north. Change in composition of the magmatic formations, if any, is associated with a reconstruction of geotectonic conditions. The second stage of magmatism manifests simultaneously in all the zones, being uniform in composition. The second stage of magmatism begins at the moment when the geotec-tonic pattern and conditions of movement are remained from the Up-per Jurassic time - the beginning of the inversione stage. Neither the surface structure nor sedimentary formations give any indica-tions of the new subsequent stage, though the composition and cha-racter of magmatism exhibit a radical change that is not a casual episode, but rather a forerunner of orogenetic state. The magmatic indicators of the endogeneous regime were evidently ahead of the tectonic events. The change in the conditions of the deep parts of the earth's crust and upper mantle manifested on the surface in variations of magma composition that reached the upper zones, as well as the tectonic moment regime.

EVOLUTION OF THE EARTH'S CRUST DURING THE ALPINE
CYCLE (GEODYNAMIC MODEL)

The available material on the tectonic and magmatic processes in the Caucasus during the Alpine cycle gives a general idea of

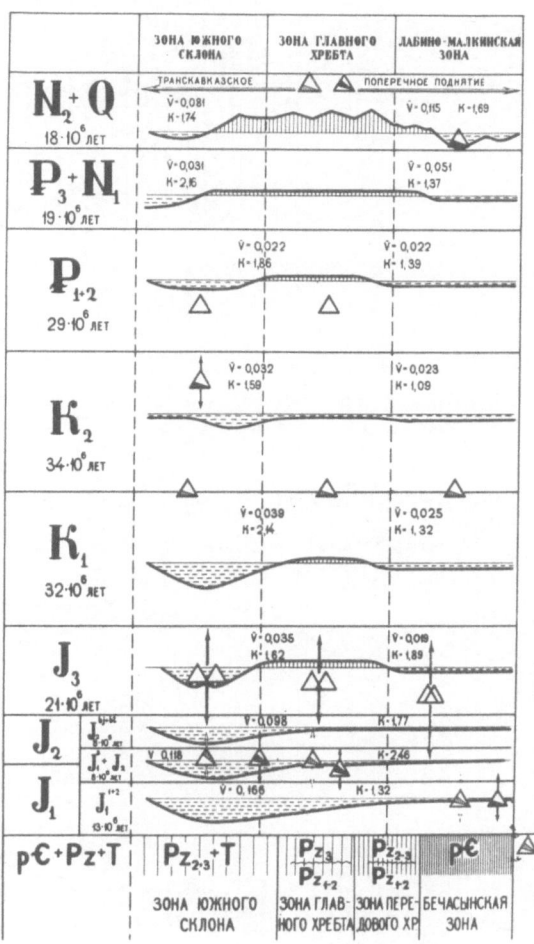

FIG. IO - Scheme of comparison of the vertical movements and magmatic phenomena during the Alpine cycle. Explanations in the text. Compiled by A.M. Borsuk and V.N. Sholpo.

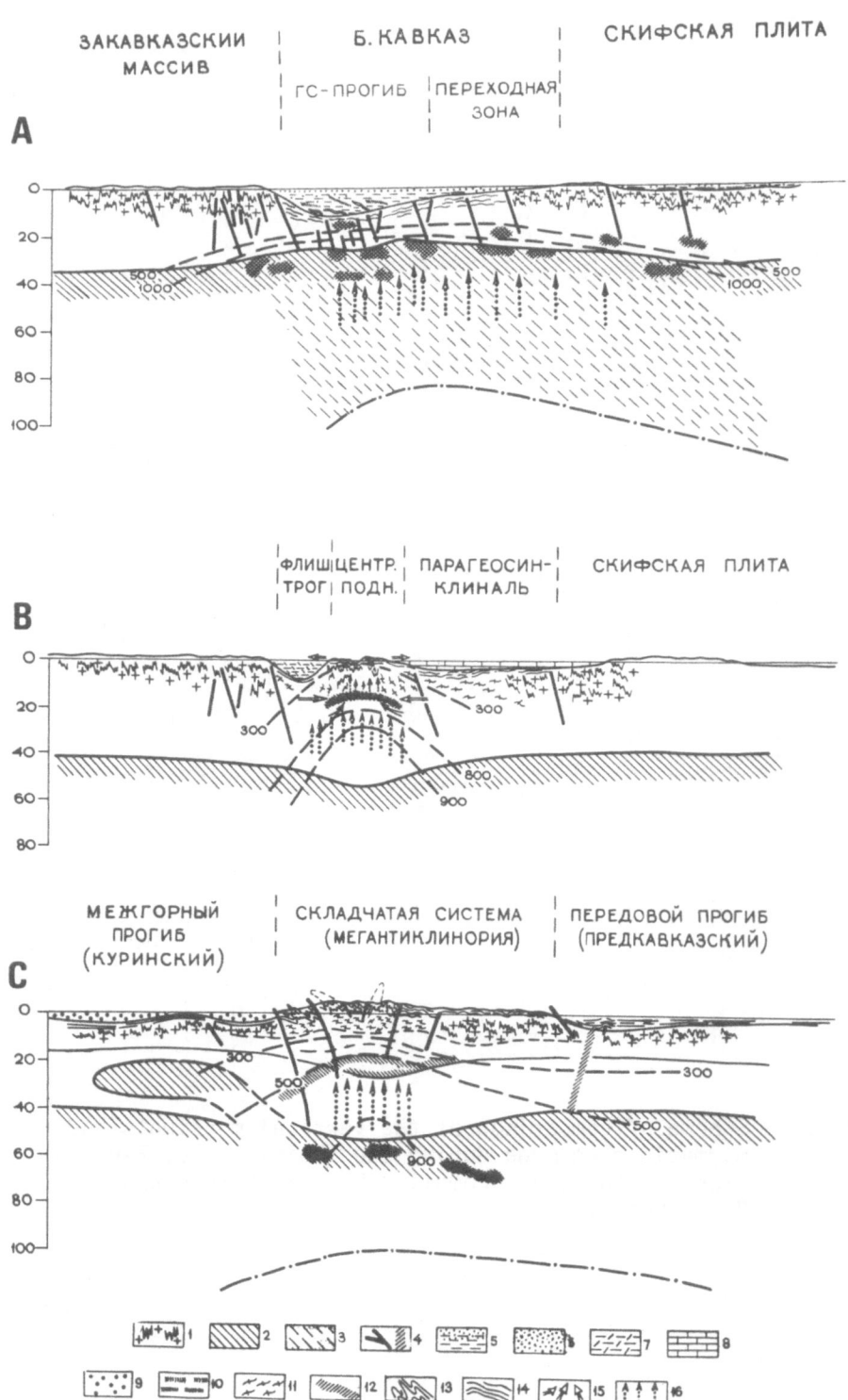

ЗАКАВКАЗСКИЙ   Б. КАВКАЗ   СКИФСКАЯ ПЛИТА
МАССИВ

ГС-ПРОГИБ ПЕРЕХОДНАЯ
ЗОНА

А

ФЛИШ ЦЕНТР. ПАРАГЕОСИН-   СКИФСКАЯ ПЛИТА
ТРОГ ПОДН. КЛИНАЛЬ

В

МЕЖГОРНЫЙ   СКЛАДЧАТАЯ СИСТЕМА   ПЕРЕДОВОЙ ПРОГИБ
ПРОГИБ   (МЕГАНТИКЛИНОРИЯ)   (ПРЕДКАВКАЗСКИЙ)
(КУРИНСКИЙ)

С

Fig. 11 - Scheme of development of the endogeneous processes in the tectonosphere of the Great Caucasus during the Alpine cycle.
A - preinversion stage; B - inversion of the geotectonic regime; C - orogenic stage
1 - "granitic" layer of the earth's crust; 2 - upper mantle; 3 - "excited" upper mantle; 4 - deep fault zones; 5 - preinversion sedimentary formations; 7 - flysch formations; 8 - parageosynclinal carbonate formations; 9 - molasse formations; I0 - sills and dykes; 11 - active granitic layer; I2 - melted zones in the crust; I3 - folded sedimentary layers; I4 - sedimentary layers with small deformations; I5 - directions of movements; I6 - traces of deep fluids.

of the character of the deep processes and makes it possible to e-
laborate the base of a geodynamic model of the evolution of the
lithosphere in the region. Data on the recent deep structure of
the area are useful in this case for introducing corrections into
the models thus obtained and assessing the probability of the inferred processes (fig. 11).

Simultaneous and interrelated manifestations of the intensive
subsidence and basic magmatism represented by gabbro-diabase formation suggest that the earth's crust at that moment - at the beginning of the Alpine cycle - was relatively thin, flexible and had
a high dispersed permeability. The fact that the magmatic formations in the mobile zone include tholeite basalt suggests that the
latter was transferred into the crust from deep subcrust horizons.
The earth's crust at this stage was relatively cold, where the
upper mantle was apparently hot.

The break of the magmatic material upward into the sedimentary units was local and occurred much later than the beginning of
the subsidence. The lateral zoning of the magmatic manifestations
in the early Alpine stage and the development in time of magmatism
(from the platform towards the mobile belt) suggest that the high
permeability zone located under the Caucasus, was not, at that time,
strictly vertical, but rather inclined, though at a fairly steep
angle, from the internal parts of the geosyncline towards the epi-
hercinian platform. This fact permits the identification of the
zone with Yu. M. Sheimann's "tectonofers" (1968).

The composition of the magmatic formations reaching the surface variated depending on the composition of the earth's crust
that deep magmatic material encountere on its way. As a result of
contamination and perhaps partially differentiation of the initial
melt intermediate intrusions of high alkalinity were formed in the
Labino-Malkinskaja zone where the initial magma passed through the
sial crust. In the transitional zone and in the axial part of the
geosynclinal trough where the earth's crust was still unstable
and, apparently, had no thick "granitic" layer, the composition of
the magmatic formations was similar to that of the initial mantle,
corresponding to the toleiite basalt.
The underwater outlows in the Transcaucasian Median Massif apparently originated from the centres localized close to the surface and,
therefore, although there might also be a thick sial crust at the
moment, it did not affected the composition of the volcanic unit
which indicates that toleiite basalt was also the initial product
for the latter.

The second phase in the development of the mobile region - inversion - was associated with the origin and development of uplifts
in the axial part of previously formed troughs. At that time granitic intrusions were formed along the axial zone, and deformation

and folding took place in the preinversion sedimentary sequence. Rocks of the lowest part of preinversion sequence underwent metamorphism of the greenschist facies stage. All events pointed to a heating of the earth's crust at that stage, as well as to associated transformation of the substance at different levels.

It is reasonable to associate the uplifting and deformation of the rocks with such physico-chemical transformations as metamorphism and granitisation. The greenshist facies of metamorphism are known to be associated with discharge of great amounts of water and a deconsolidation of the matter. This accounts for inversion of densities at various horizons of the earth's crust, i.e. the state of instable equilibrium which may cause the diapir-like arising and intrusion of deconsolidated masses.

At that stage the earth's crust decreased in permeability. Granitoid intrusions that originated at the very beginning of the inversion stage made the crust virtually impermeable, apparently increasing its thickness and making it more rigid.

There is no evidence of any magma activity at this stage, allowing the assumption that after giving the energetic impact to the earth's crust in the first stage, the mantle had cooled by that time. So, if in the first stage of the preinversion subsidence the mantle was hot, excited and active, it became rather cool in the second stage. At the same time, the earth's crust being relatively cold in the first stage was heated and became active in the second stage.

A new phase in the development of the Greater Caucasus started in the Neogene-Quaternary orogenetic stage. Typical of that stage was transverse anticaucasian division of the Greater Caucasus into blocks, the unequal relative displacement of which made pominent transverse uplifts and subsidence of different orders. The superimposed nature of the movements is emphasized by the specific character of magmatism at that stage, which manifested simultaneously and monotonously in all structural-historical zones of the Greater Caucasus and far beyond. Indications of the change of the movement regime were prominent in magmatism long before the Neogene-Quaternary period, at the boundary between Early and Late Cretaceous. Even then, however, the first portion of a new deep melt penetrated from the depths to give rise to the high alkaline essexite-teschenite formation.

As indicated by magmatic manifestations of the early stage (Cretaceous) and subsequent ones (Neogene-Quaternary) as well as by geophysical evidence on the conditions of the deep horizons of the tectonosphere, the new thermal impulse causing the Neogene-Quaternary orogenic movements was delivered to the tectonosphere and, probably, to the base of the earth's crust along a nearly north-south tending zone. The energy from the excited astenosphere was transferred in a transverse direction before the beginning of the orogenic stage. This may explain the nature of orogenic movements in the Greater Caucasus.

The composition of the magmatic material supplied at that time from the astenosphere to the earth's crust was not hardly different from that transferred upward during the first stage of the Alpine cycle.

In any case, there is no grounds for presuming that the composition of the melt from the astenosphere could change greatly. In this case, however, the deep matter encountered the earth's crust with a different structure and composition: thicker and more rigid, without dispersed permeability after granitisation and metemorphism. Accordingly there was a different reaction by the earth's crust to the similar energetic impulse. At this time the crust bulged up, being "pushed" by asthenoliths rising from the upper mantle, and broke down into blocks. All this prompts the quite definite conclusion that the orogenic stage was not a direct continuation of the geosynclinal cycle; it was an independent event and it was only in the Caucasus that it superimposed over an area that had just finished its geosynclinal evolution.

REFERENCES

Adamia Sh.A., Zakariadze G.S. and Lordkipanidze M.B., 1977, The evolution of the ancient active continental margin on the example of the Caucasus history, Geotectonics, 4, 11-32.
Adamia Sh.A., Lordkipanidze M.B. and Zakariadze G.S., 1980, On the problem of the Tethys ocean (by data for the Caucasus and adjoining regions), 26e Congrès Gèologique Int., Abstracts, Paris, 310.
Artemjev M.E., 1966, The isostasy anomalies and some aspects of their geological interpretation. M., Nauka, 138 (in Russian).
Artemjev M.E., 1971, Some features of the subsurface structure of Mediterranean-type basins, according to the isostasy anomalies data, Bull. MIOIP, XLVI, 4, 39-52 (in Russian).
Artemjev M.E. and Balavadze B.K., 1973, Isostasy of the Caucasus, Geotectonics, 6, 20-33.
Beloussov V.V., 1978, Endogeneous regimes of continents. M., Nedra, 232 (in Russian).
Beloussov V.V., 1981, Endogeneous regimes of continents. M., MIR
Borsuk A.M., 1974, Igneous formations in the evolution of a geosynclinal system (on the example of Alpine areas of the Great Caucasus), In "Actual problems of modern petrography," M., Nauka, 163-171 (in Russian).
Borsuk A.M., 1977, Igneous formations as indicators of the endogene-

ous regime of a mobile area, Izv. AN USSR, ser. geol., 2
    73-85 (in Russian).

Dewey J.F., Pitman W.C., Ryan W.B.F. and Bonnin J., 1973, Plate
    tectonics and the evolution of the Alpine system, Bull. Geol.
    Soc. Am., 84, 3137-3180.

Gamkrelidze I.P., 1976, The formation mechanism of the tectonic
    structures and some general problems of tectonogenesis,
    Tiblissi, Metsniereba, 226 (in Russian).

Khain V.E., 1973, General geotectonics, M., Nedra, 512 (in Russian)

Khain V.E., 1975, The main stages of the tectono-magmatic develop-
    ment of the Caucasus: a study for geodynamic interpretation,
    Geotectonics, 1, 13-27.

Leonov M.G., 1975, Wildflysch of the Alpine areas. M., Nauka, 140
    (in Russian).

Lubimova E.A., Polyak B.G., Smirnov J.B., Sergienko S.I., Koperbach
    E.B., Lyusova L.N., Kutas R.I., Firsov F.V., 1973, A review of
    the heat-flow data for the USSR. In"Heat flow from earth's
    crust and upper mantle", M., Nauka, 154-195 (in Russian).

Malovitskiy J.P., 1970, On the main submeridional dislocations in
    the southern European part of the USSR, Geotectonics, 3,
    115-122.

Sheimann Yu. M., 1968, Studies of subsurficial geology (on the rela-
    tion between tectonics and magma formation). M., Nedra, 231
    (in Russian).

Sholpo V.N., 1962, On folding floors in the Shah-Dag zone, South-
    east Caucasus, In "Folding of the earth's crust.."M., Nauka,
    199-218 (in Russian).

Sholpo V.N., 1964, Folding types and conditions in Slantzeviy Dagh-
    estan. M., Nauka, 167 (in Russian).

Sorskiy A.A., 1966, Major features of the structure and development
    of the Caucasus in the relation of its deep setting, In"Deep
    structure of the Caucasus", M., Nauka, 22-34 (in Russian).

Vinnik L.P., Lenartovitch E., 1976, The upper mantle structure of
    the Caucasus and Carpathians, according to the seismic data,
    Phys. Earth, 3, 3-14.

SEISMICITY AND CRUSTAL STRUCTURE IN THE

ITALIAN REGION: A PRELIMINARY ZONING

Roberto Cassinis

Istituto di Geofisica
Università di Milano
via G.B. Viotti 3/5, 20133 Milano

SUMMARY

The transitional area of Italy is a very complex one, the relationship between seismicity and crustal structure being complicated
by the geological history of the more general collision between
Eurasian and African plates.

The final objective of the investigation is to analyze the elements of instability in order to use them as weighted parameters for
the definition and classification of the earthquake origin zones.

The phase of the survey illustrated in this paper consists in
the preparation of a preliminary zoning, in the determination of the
average values of the available parameters for each zone and in the
comparison between the major units as far as the character of the
seismicity and the structural type is concerned.

1. INTRODUCTION

During the last years, chiefly owing to the activity of the
national program in Geodynamics, a larger amount of new data has
been collected and historical data have been revised on the seismicity
of the Italian area; moreover, several new surveys have been carried
out on crustal structure using geophysical methods and especially
deep seismic soundings.[1,2] Furthermore, new comprehensive data have
been gathered by the national Agency for oil (ENI-AGIP) which are of
general interest, as the aeromagnetic map of Italy and adjacent seas[3]
(only some sheets have been published so far) and the results of the
recent seismic reflection surveys on the Po Valley.[4]

The final objective of our investigation is a quantitative analysis of the parameters that can be relevant with the instability of the Crust. Their values have to be properly weighted and statistically compared for the definition and classification of the earthquake origin zones.

The present study consists in the preparation of a preliminary zoning based mainly on seismic historical data and in the comparison of the different zones. The parameters that have been considered so far are those that are available on the whole area, although their reliability and density often vary from zone to zone to a considerable extent.

2. AVAILABLE DATA ON SEISMICITY

- ENEL (Ente Nazionale per l'Energia Elettrica) general catalogue of Italian earthquakes for the period 1000 - 1975 A.D. (20,580 events);
- Seismological bulletins issued by I.N.G. (Istituto Nazionale di Geofisica) with the data of the national seismographic network I.N.G., C.N.R.
- Progetto Geodinamica, collected by an average of 40 short period observations during the period 1976 - 1980 (1,946 events). A preliminary filtering of the aftershocks has been applied to the data.

A Fortran IV program has been prepared for the extraction of data both from the catalogue and from the bulletins that have been transferred on CCT's. A base grid with an elementary cell of 10' (ϕ) x 15' (λ) has been used. The size of the cell is compatible with the accuracy of the zone boundaries; furthermore, it seems a good average standard for the statistical analysis of the majority of the parameters. The geographical grid is based on the Lambert conical projection.

Epicentral maps (Figure 1) have been prepared for the whole region (scale 1:1,500,000) and for particular regions (scale 1:500,000), separating the events into three periods:

    1000 - 1890 (about 5000 events)
    1891 - 1975 (about 15000 events), and
    1976 - 1980 (1,946 events).

The epicenters are classified according to the magnitude. For the two historical periods the magnitude of the majority of the events has been obtained from the I_o using the Karnik relationships:[5]

$$M_{(1)} = 0.5 \times I_o + \log_{10} \underline{h} + 0.35$$

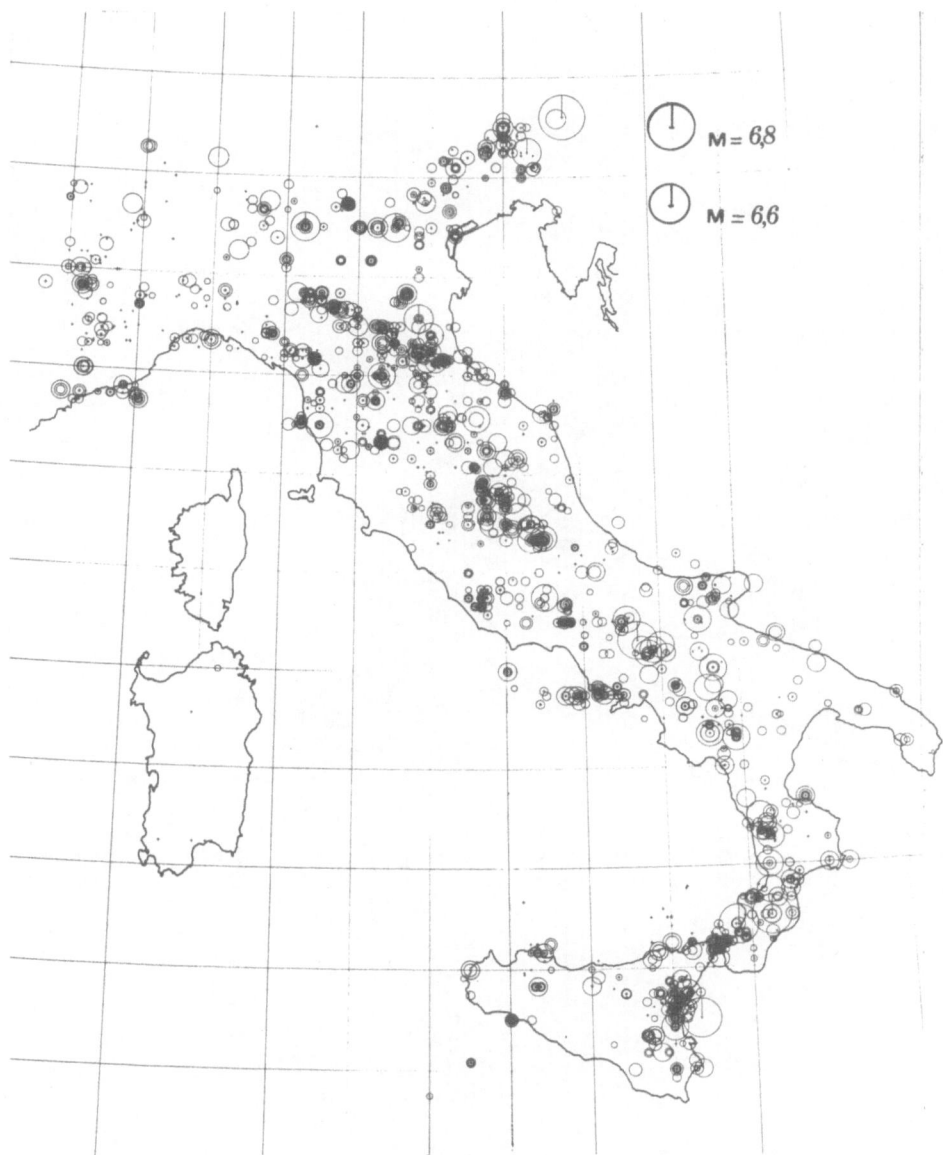

Fig. 1 a. Epicenters of the period 1000–1890 A.D. Crosses:
magnitude not indicated.

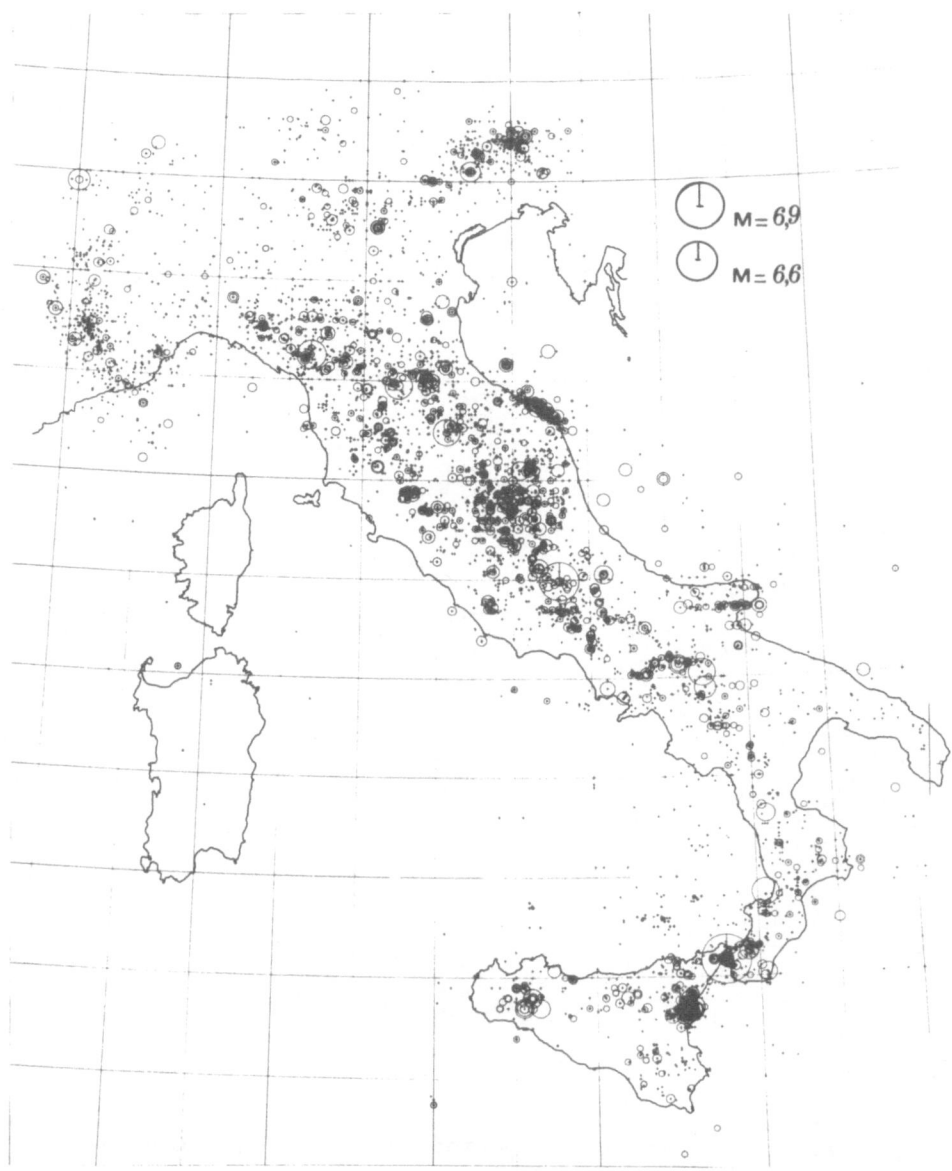

Fig. 1 b. Epicenters of the period 1890–1975. Crosses:
magnitude not indicated.

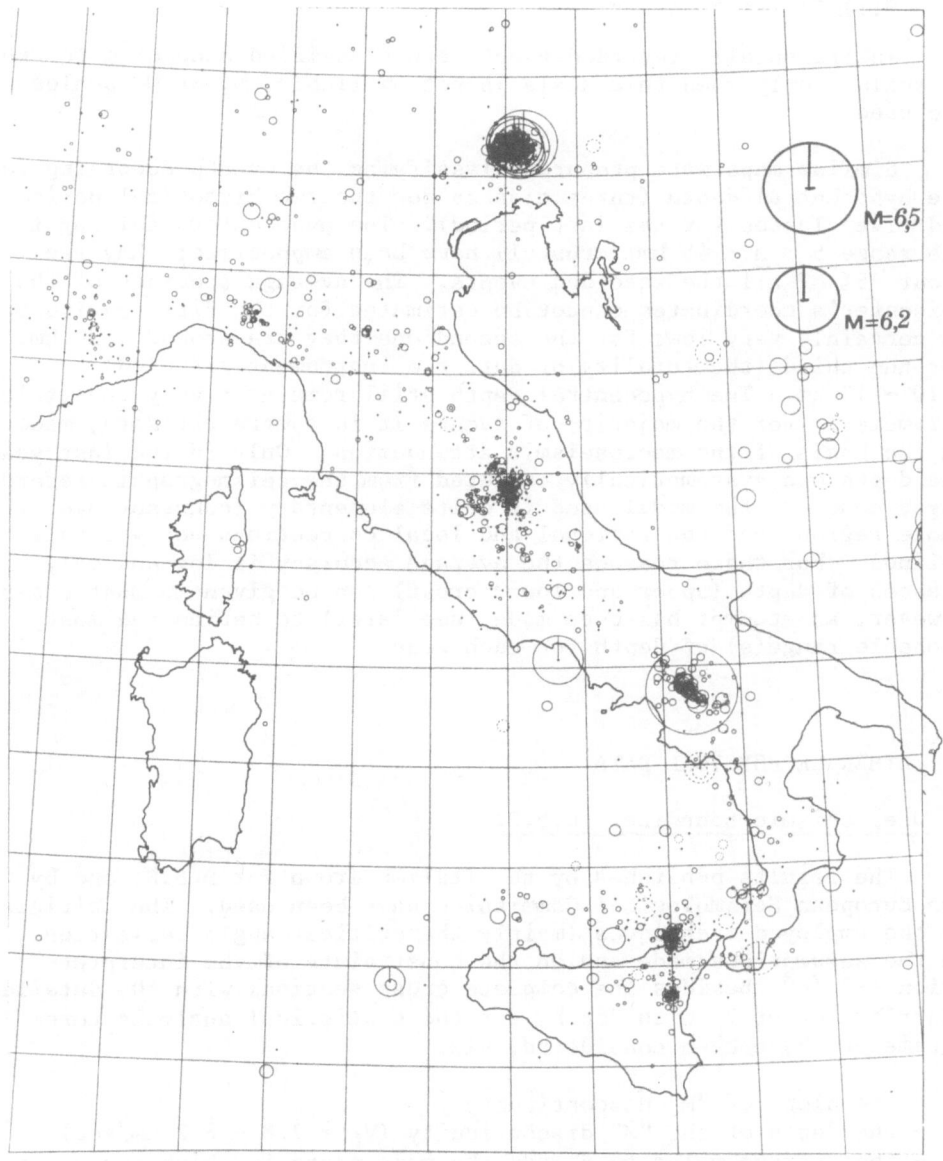

Fig. 1 c. Epicenters of the period 1976–1980 (the aftershocks are unfiltered).

where $\underline{h}$ is the hypocentral depth or, if $\underline{h}$ is not expressed,

$$M_{(1)} = 0.53 \times 1_o + 0.95$$

Instrumentally recorded events are classified according to the M_L scale: only when this scale is not available, M_s or M_b scales are used.

Similar maps were prepared classifying the events according to the hypocentral depth (three classes for the two historical periods and five classes for the last period). The hypocenters falling in the range $5 < \underline{h} < 45$ km (crustal) have been emphasized; they are about 95% of all the recorded events. The average accuracy of the epicenter's coordinates cannot be estimated for the first period but is certainly very low; for the second one they are around ± 30 km, for the third (the totality of data are instrument gathered) ± 10 - 15 km. The hypocentral depth still remains a very uncertain parameter: for the majority of events it is a mere estimate, made on the basis of the macroseismic attenuation. Only in the last years the depth was systematically computed from the seismographic recordings; however, the model used is quite elementary (constant over the whole region) and the regional and local corrections not yet well defined. For these reasons the average accuracy is low and only classes of depth (upper and lower crust) can be given in most cases. However, an attempt has been made (see later) to define the most probable range(s) of depth for each zone.

3. OTHER GEOPHYSICAL DATA

a) Deep seismic soundings (D.S.S.)

The results published by the Italian Group for D.S.S. and by the European Seismological Commission have been used. The criticisms on the employed techniques (mainly the critical angle reflection), on the assumptions made and on the constraints of the interpretation.[1,6,7,8] Besides the complete cross sections with the detailed distribution of V_p with depth, for the statistical analysis three parameters have been considered, viz:

- the slope of "M" discontinuity
- the depth of the "M" discontinuity ($V_p = 7.8 - 8.2$ km/sec)
- the average velocity V_p for the main parts in which the crust can be seismically divided, viz, the sediments, the upper crust and the lower crust.

The depths of three boundaries have been indicated, the first where V_p reaches 5.5 km/sec, the second for $V_p \geqslant 6.7$ km/sec, the third where the max velocity gradient in the lower crust is met.

b) Gravity data

The only available comprehensive map of Bouguer anomalies is, so far, the one published by the Italian Geodetic Commission in 1972,[9] in the scale of 1:1,000,000. Also the isostatic anomalies are available but the coverage is not complete and the assumptions made for the calculations are no longer compatible with the known structure of the Crust and Upper Mantle.

c) Aeromagnetic map of Italy

The published comprehensive map,[3] gives but a very poor and scanty outline of the magnetic lineaments of the region, while the high accuracy of the survey and of the data processing would allow a very good analysis. Nevertheless, some striking features are clear; the transition from the Tyrrhenian domain to the Adriatic foreland, the absence of both shallow and deep sources along the Apenninic chain; an alignment of deep seated sources at the border of the Adriatic foreland; two broad and weak anomalies along the axis of the Po foretrough the most western of which has been found to correspond to volcanic intrusions.

4. GEOLOGY AND TECTONICS

The "tectonic map" of Italy (scale 1:1,500,000) prepared by the "Progetto per la Geodinamica" of C.N.R. has been assumed as a reference map. This document is the result of the work of many authors (see, for general comments references[10,11,12]). The main tectonic lineaments, some indication of sub-surface geophysically outlined structures, and the lithology of the major geological units are shown. Mainly on the bases of this map, a preliminary subdivision of the region into 19 zones has been attempted according to the main lithologic and tectonic features (Figure 2). As it will be shown later, only some zones correspond to the sub-surface and crustal structure, owing to the masking produced by the mobility of the overburden and by the fluctuations of several orogenic "waves". It is well known that Italy is one of the most complex transitional areas in the world: several orogenic phases, involving change of stress direction, have occurred along a relatively short time span. The lifting of the two chains, the Alps and the Apennines, is not yet completed; the area is characterized by very large horizontal displacements of folds and nappes, the basins being filled by thick detrital deposits and by submarine slides. Because of the allocthonous blankets of highly variable thickness the subsurface geology is masked; when the overburden is composed of "flysch" or "argille scagliose", as happens along the majority of the Apenninic region, the geophysical exploration becomes very difficult; even the most sophisticated techniques of seismic reflection are often uneffective

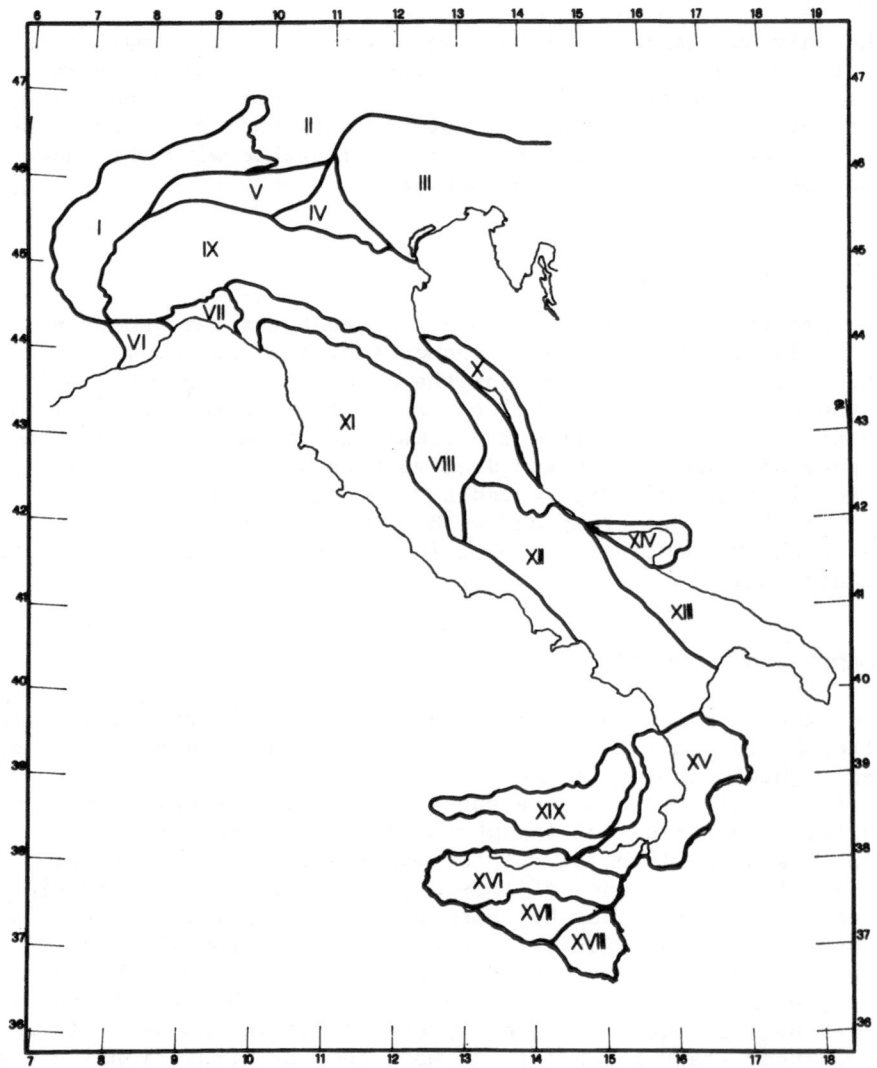

Fig. 2. Geological units after the geological and tectonic map.
I - Pennides (part). II - Austrides (part). III - South
Alps (E). IV - Garda. V - South Alps (W). VI - West
Liqurides. VII - Transitional zone. VIII - North and
Central Apennines. IX - Padanian and Adriatic foredeep.
X - Conero. XI - Tuscany, Latium, Campania. XII - South
Apennines. XIII - Apulia foreland. XIV - Gargano. XV -
Calabro-Peloritan arc. XVI - North Sicily. XVII - Caltani-
setta trough. XVIII - Ibleian foreland. XIX - Aeolian arc.

in giving information on the "autochtonous" formations when the thickness of the clastic overburden reaches 1 km or even less. Also the interpretation of gravity anomalies is hampered by the strong lateral density and thickness variations of the overburden.

Even if there is little accurate information on the hypocentral depth, it is clear that the earthquake sources are inside the Crust, except the deep ones along the "Aeolian arc". But, owing to the superposition of different tectonics and to the lateral variations in rigidity, the "seismo-active layer" in some regions is in the upper crust (sometimes in the sedimentary overburden), while in others it is in the lower crust; both upper and lower crust can be active in some cases. A reliable determination of the focal mechanism would help considerably. However, so far, only in a very few cases can the mechanism be considered acceptable. There are several reasons for this; the scarcity of good instrumental data on strong or moderate earthquakes; the complexity of the mechanism itself, that is very seldom of pure transcurrence; the strong lateral variations which make the instrumental observation more difficult. During the last years the seismographic networks have accumulated a good deal of new data, but they concern small events or after-shocks that have migrated to uppermost layers, while the main shocks have been studied mainly on the basis of the pattern of isoseismals.

The generally accepted geodynamic model is of the type shown in Figure 3 where the region is subdivided into units that exhibited a homogeneous behavior during the last 10 My. The main hypotheses supporting this model are the expansion of the North Tyrrhenian (Ligurian sea) and of the Tuscan crust; the expansion of the South Tyrrhenian and the general counterclockwise rotation of the Peninsula; the resulting closure of the Adriatic sea, the boundary between the distension and the compression areas corresponding, in a broad sense, to the dividing line of the Apenninic chain.

5.   CRUSTAL STRUCTURE FROM D.S.S. RESULTS

The debate is still open on the extent and existence of the velocity inversions, on the shape of the velocity functions and on the type of the transitions from the upper to the lower crust and from the lower crust to the upper mantle. Notwithstanding that the existence of a seismic discontinuity at the base of the crust is beyond question. It is only remarked that in some areas of the continental crust, the "M" transition is less sharp and becomes "complex".

It has been demonstrated that, in most cases, the depth of the maximum gradient around 8.0 km/sec does not change very much using different approaches for the interpretation.[6] For this reason we considered as the most significant parameters the slope or, better, the lateral discontinuity, and the depth of the "M" transition. As

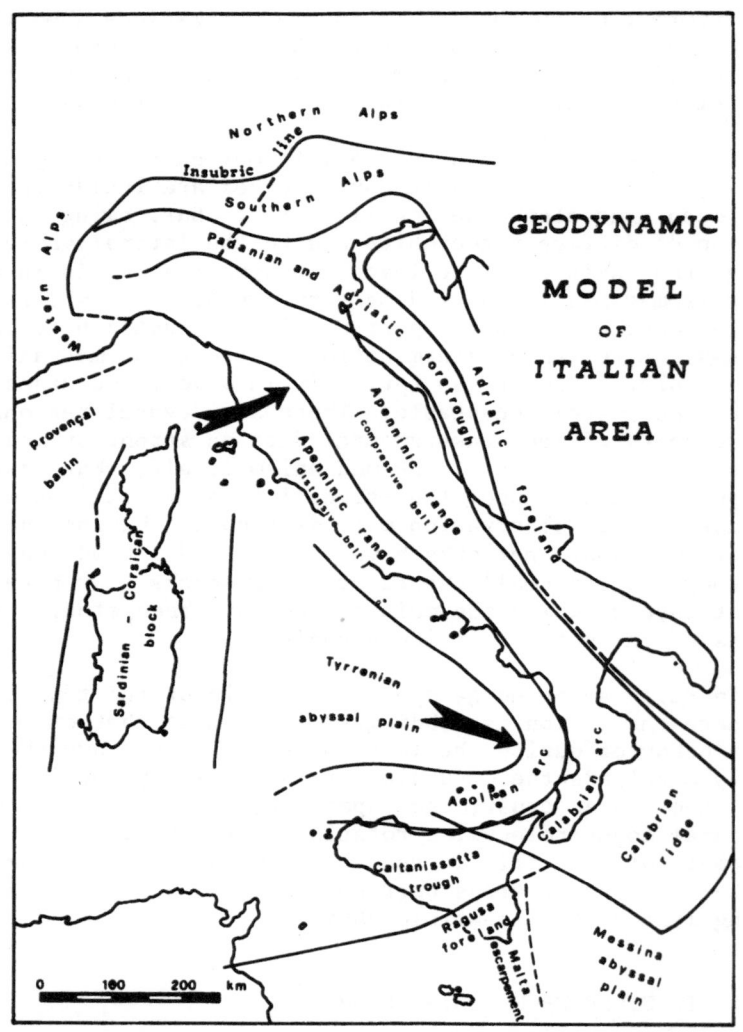

Fig. 3.   Geodynamic model of Italy (after Praturlon and Scandone, 1977).   Each unit is supposed to have experienced a homogenous behavior during the last 10 My.

second order parameter, we assumed the average velocities of P waves as stated in section 3a. The major feature shown by the D.S.S. is a very strong slope of the "M" that has been found along several profiles crossing the axis both of the Alps and of the Apennines. As examples of this feature, the data and interpretation of two typical D.S.S. cross sections are shown in Figures 4 and 5. The position of the profiles is indicated in the map of Figure 6. The hypocenters found on a strip 30 km wide (extending 15 km on both

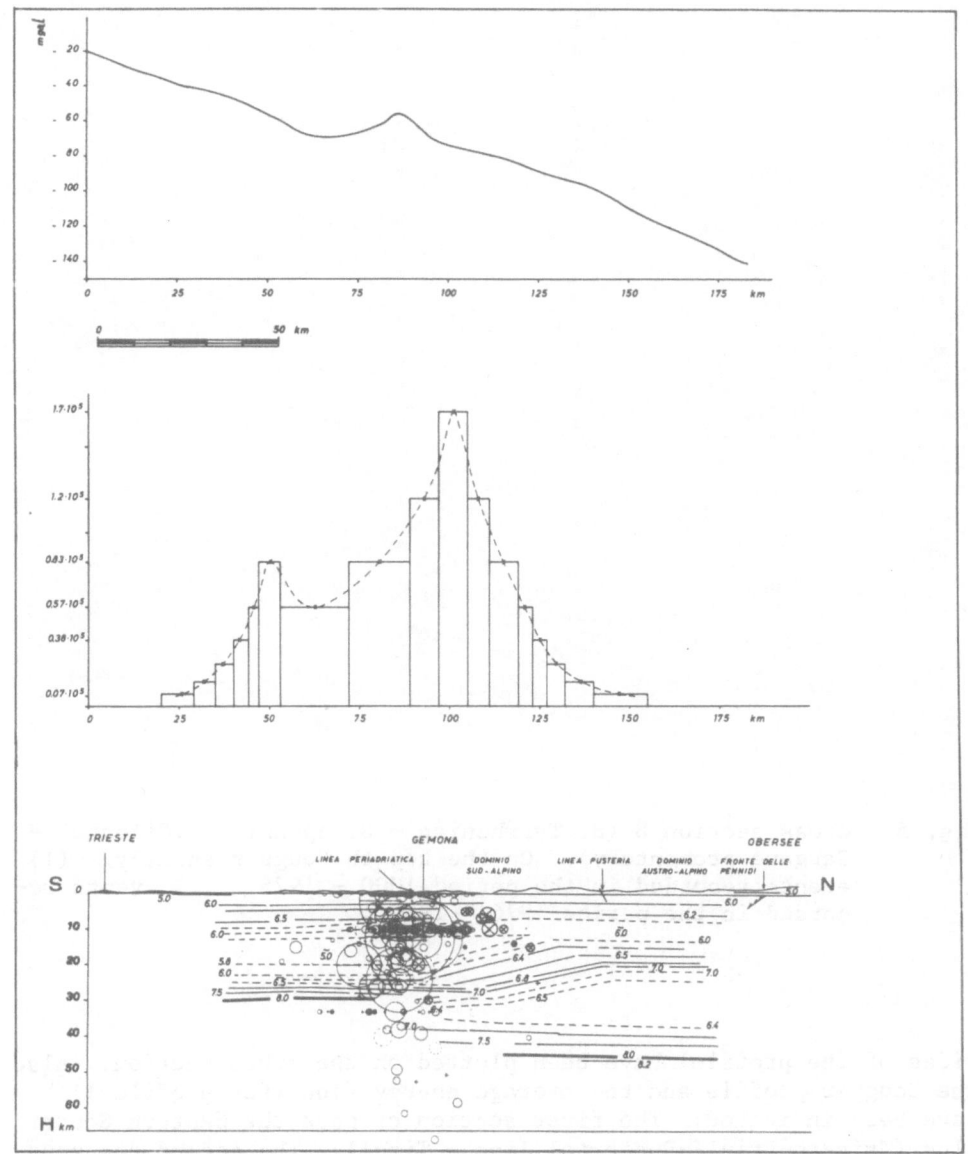

Fig. 4. Cross section A Trieste-Tirol (see for the position the map of Figure 6). From the top: profile of Bouguer anomalies; "tectonic flow" (for the period 1000 - 1975) in $\text{Erg}^{1/2}/\text{km}^2.\text{y}$; interpretation of the D.S.S. profile with the indication of P velocities and of the projection of hypocenters falling within 15 km on both sides of the section.

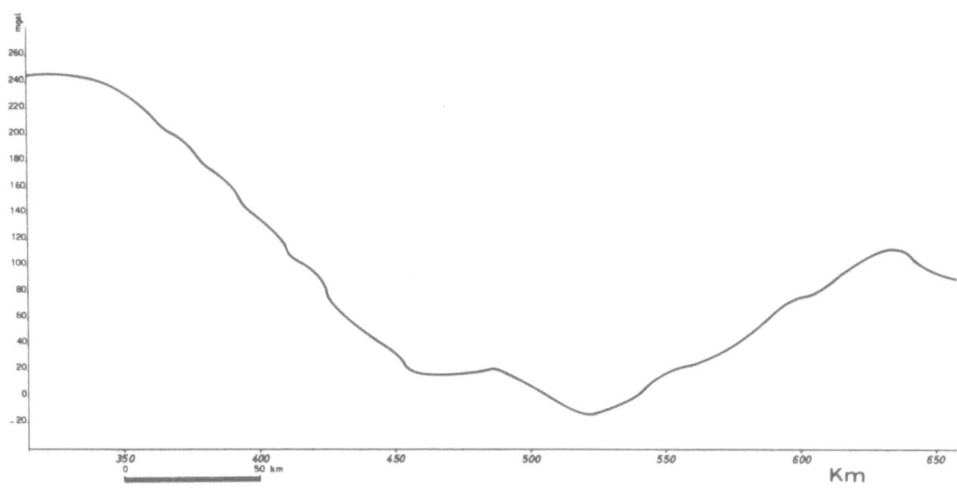

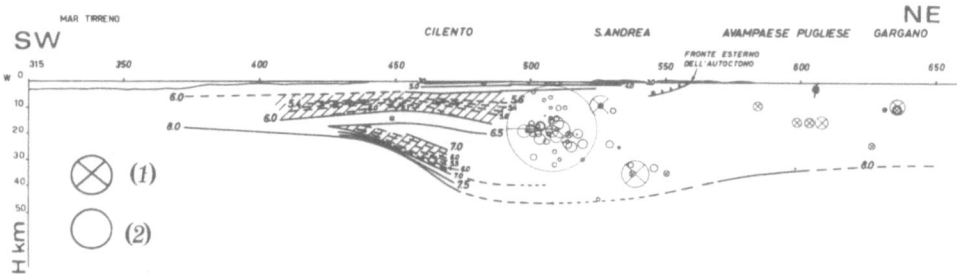

Fig. 5. Cross section B (S. Tyrrhenian - S. Apennines (Cilento) - Gargano promontory). On the top th Bouguer anomaly. (1) events recorded in the period 1000 - 1975. (2) events recorded in the period 1976 - 1980.

sides of the profile) have been plotted on the cross section. Also the Bouguer profile and the average energy flow (for profile 1)[20] have been indicated. The first section crosses the Eastern South Alps (Trieste-Friuli-Pustertal line - Tirol). The second one runs from the South Tyrrhenian coast of Cilento (S. of Naples) to the S. Apennines (Irpinia) and to the Gargano promontory. In both cases the D.S.S. show a highly variable structure of the crust that can be interpreted like a "jump" of the "M" or an overthrust of two crustal domains or even a mixed zone of crustal and upper mantle material according to the "flake tectonics" hypothesis.[13] The seismic activity is clearly inside the crust but a correlation between the mechanical properties (rigidity) and the foci is very uncertain. It is clear, however, that in both cases the seismicity concentrates in

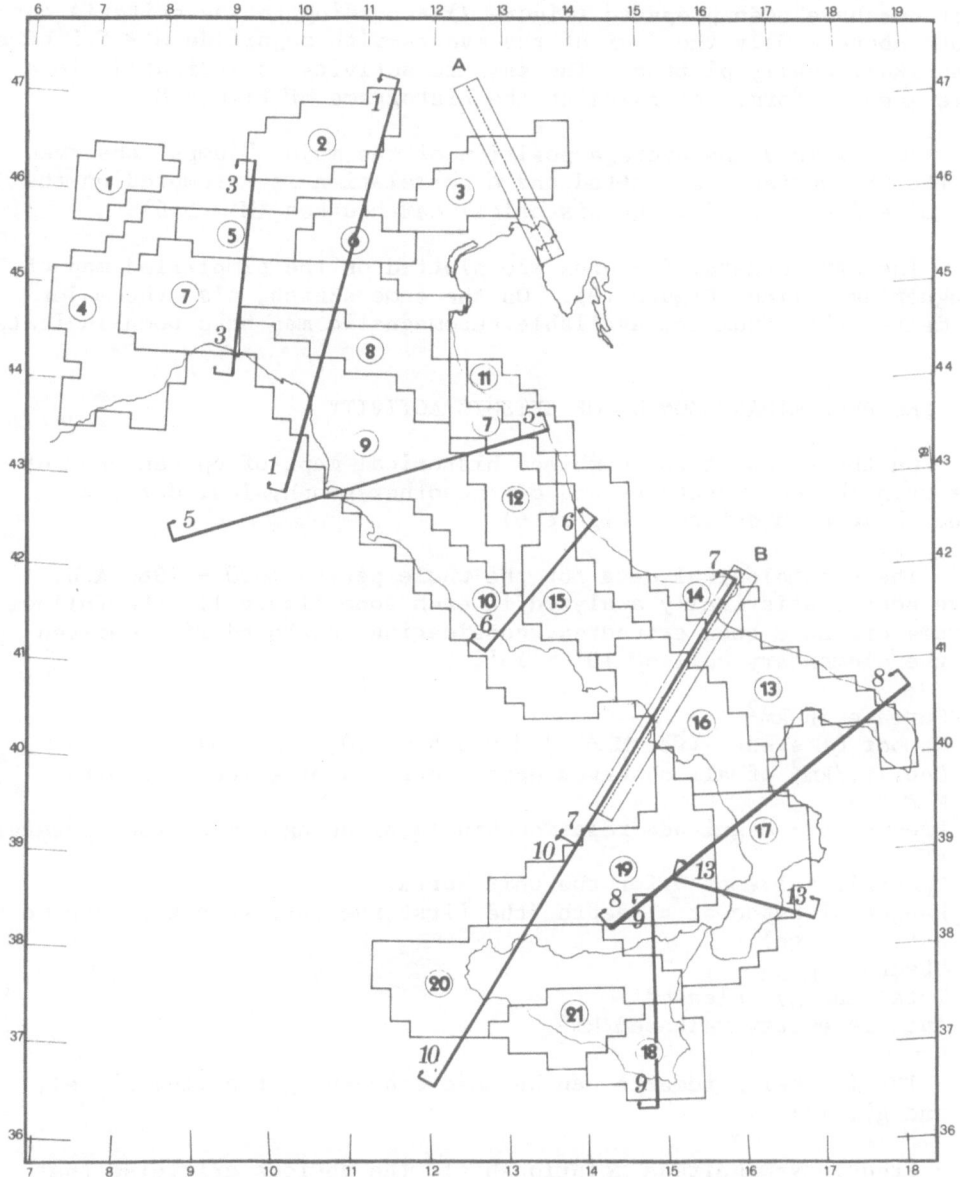

Fig. 6. Preliminary zoning considering the seismological activity.
The traces of the cross-sections of Figure 7 are also
shown.

65

the upper edge of the structural discontinuity (the "Adriatic" and the "Tyrrhenian" crust). In order to compare the seismicity to the structure in a more comprehensive and generalized form several cross sections have been prepared (Figure 7) according to the criteria outlined above. Only the foci of the events with magnitude M $\geqslant$ 5.5 have been individually plotted. The seismic activity is indicated, in a more general form, according to the histograms of Figure 8.

In Figure 9 the average position of the major "jumps" observed on the "M" surface is plotted and a correlation is attempted on the map of epicenters of large historical earthquakes (M $\geqslant$ 5.5).

The same crustal features are plotted on the simplified map of Bouguer anomalies (Figure 10). On the same sketch, also the major elements taken from the available aeromagnetic map have been indicated.

6. THE PRELIMINARY ZONING OF SEISMIC ACTIVITY

On the bases of the combined historical maps of epicenters, of the main tectonic features and of the other geophysical data, 21 zones have been defined (Figure 6).

The seismological data for the whole period 1000 - 1980 A.D. have been statistically analyzed in each zone (Table 1); the following parameters have been extracted, considering the boundaries defined by the elementary cell of 10' x 15':

a) Surface in km^2
b) Number of events (for M $\geqslant$ 6, M = 5.5 - 6.0, M $\geqslant$ 3.5)
c) Density/km^2 of all observed epicenters and of epicenters with M $\geqslant$ 3.5
d) Frequency - magnitude relationship (a,b) using a magnitude interval $\Delta M = 0.2$
e) Specific value of a for the unit surface
f) Ranges of hypocentral depth (the first row corresponds to the most active range)
g) Observed M_{max}
h) Total energy released
i) Average energy released/km^2.

The following remarks can be made concerning the item d), e), f) and g), i):

- Frequency-magnitude relationship: the obvious criticism that can be raised concerns the length of the considered period and the M_{min} (3.5), that is the use of inhomogeneous data. On the other hand, a low threshold and a long period mean a sufficient number of events for a reliable use of the regression analysis. The values of b range between 0.59 and 1.11, their average being around 0.72. If four

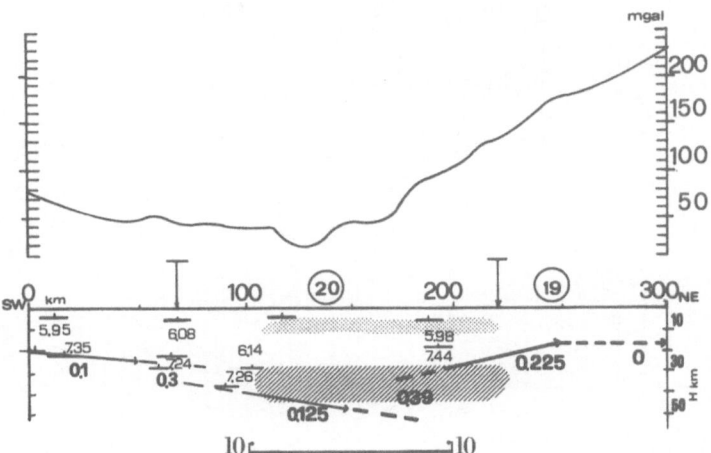

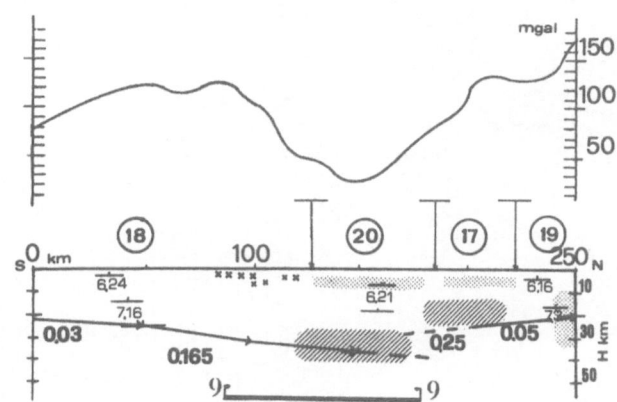

Fig. 7. Schematic sections across the Alpine and Apenninic chain
(for position see Figure 6). Explanation of symbols in
the following pages.

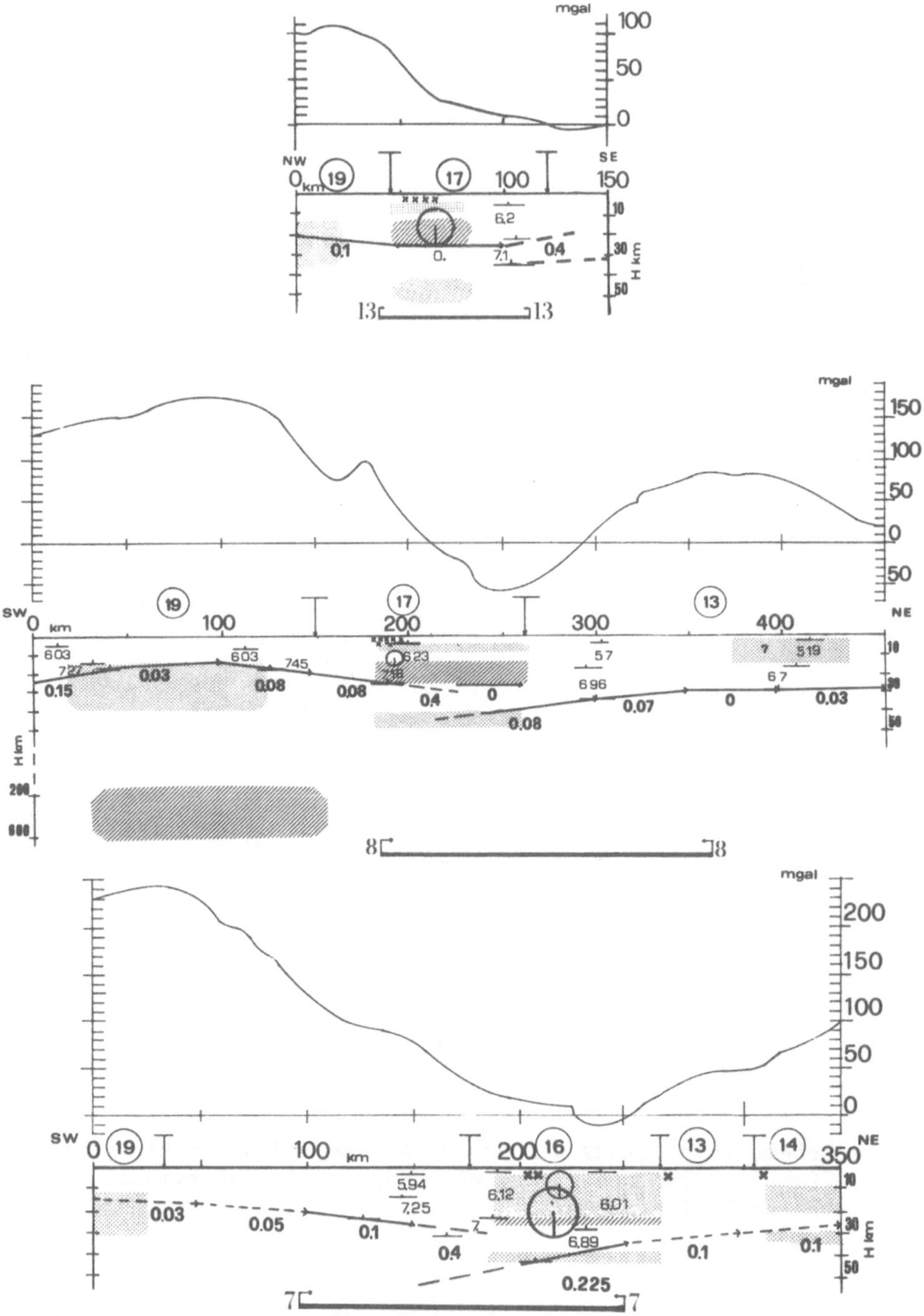

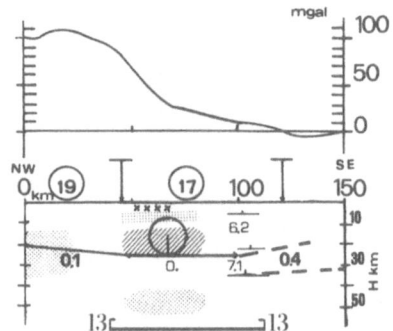

Fig. 7. (continued)

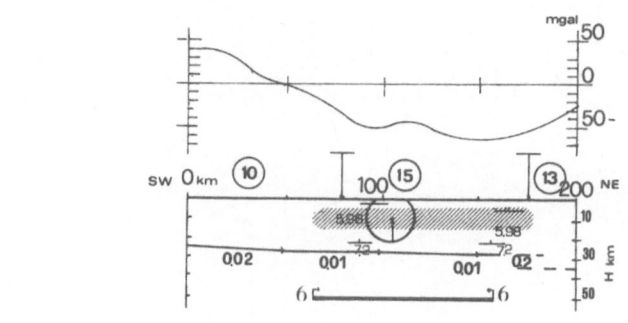

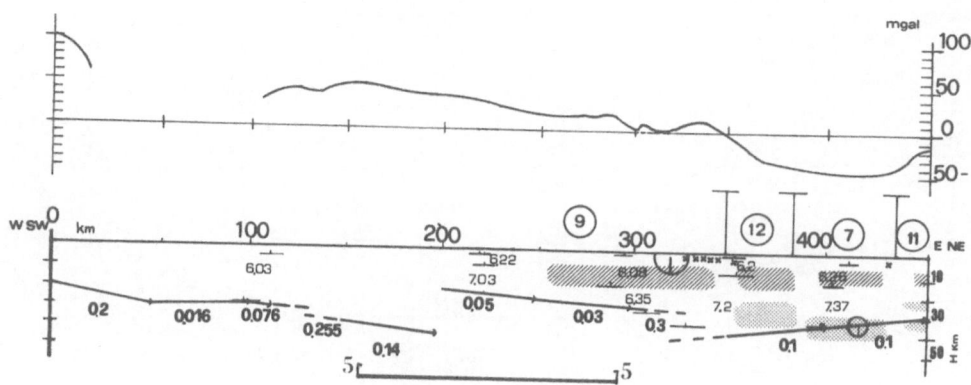

Fig. 7. (continued)

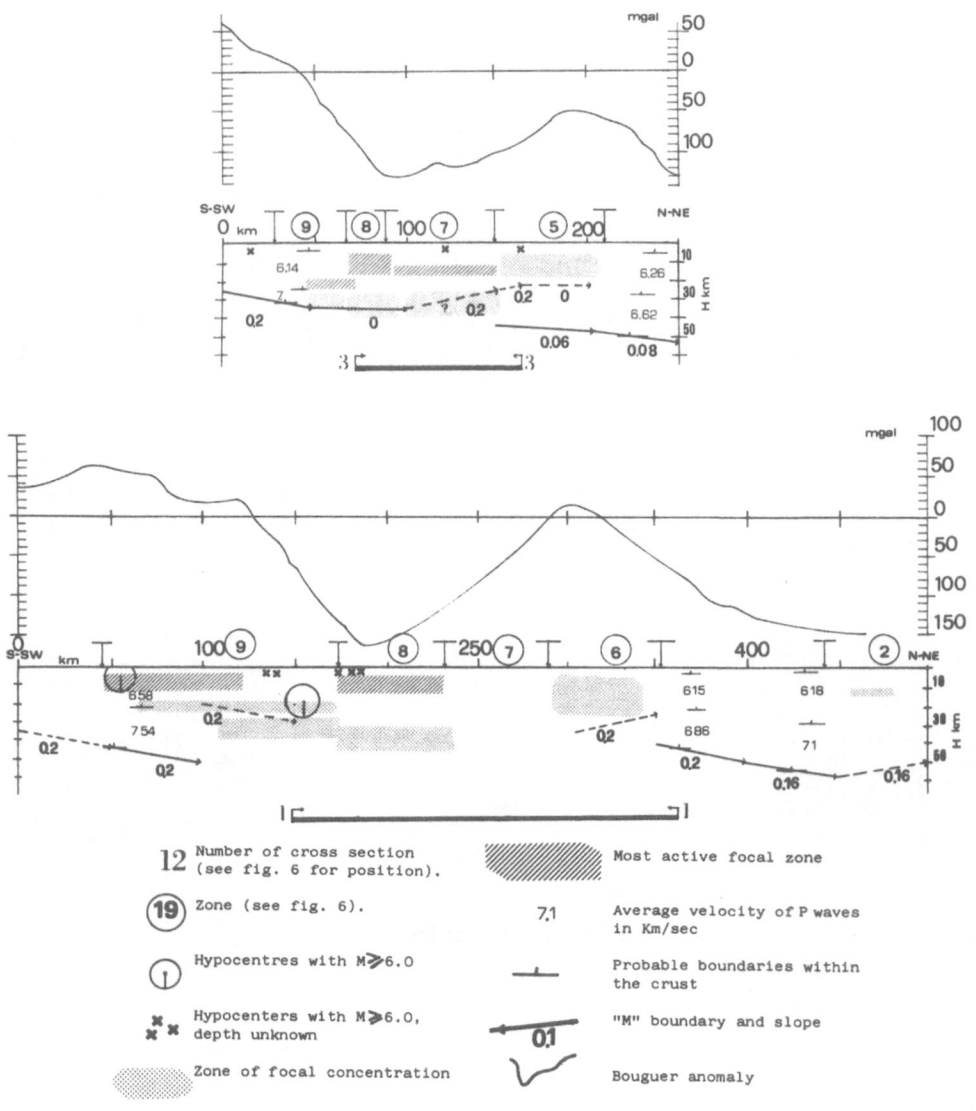

Fig. 7. (continued)

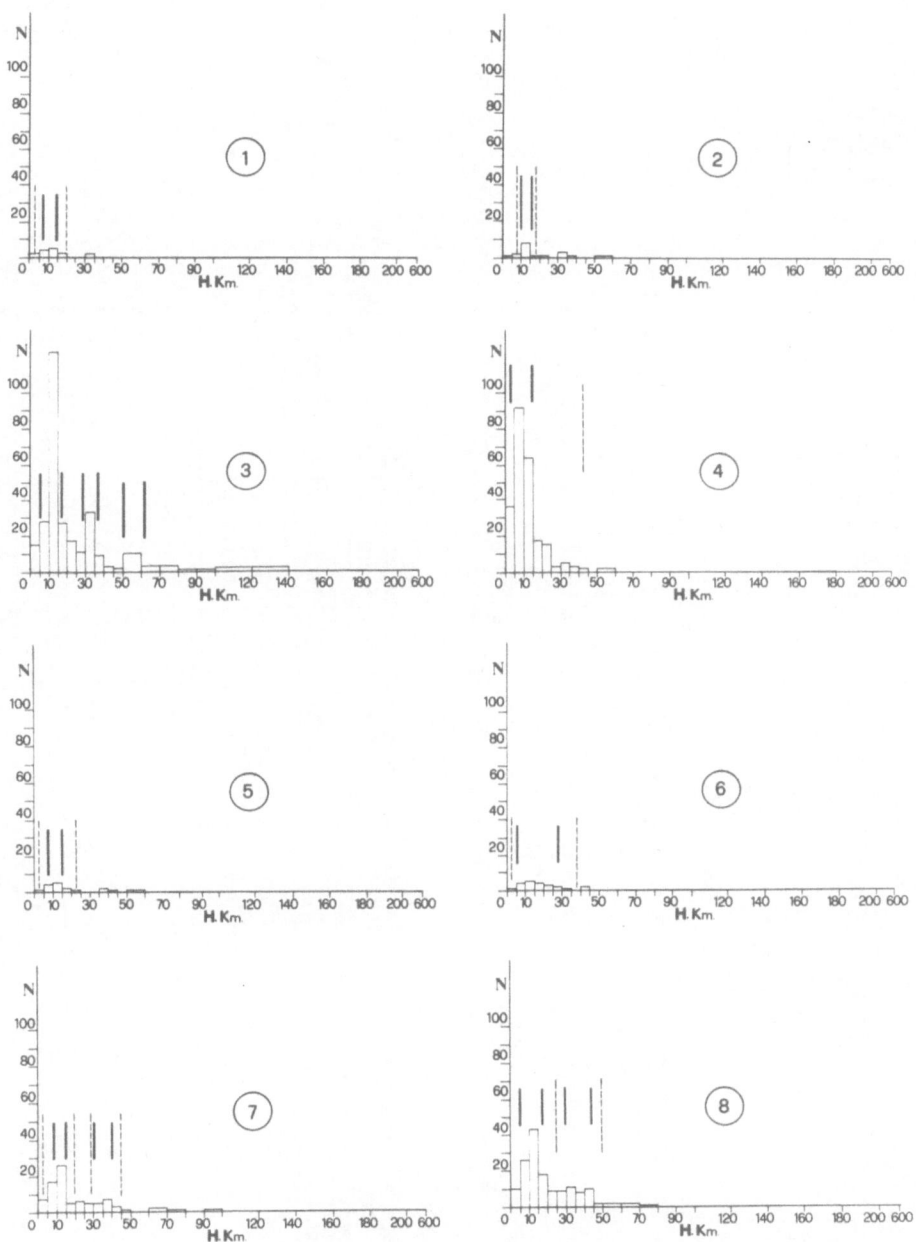

Fig. 8. Histograms of focal depth for the zones of Figure 6 (see text for explanation).

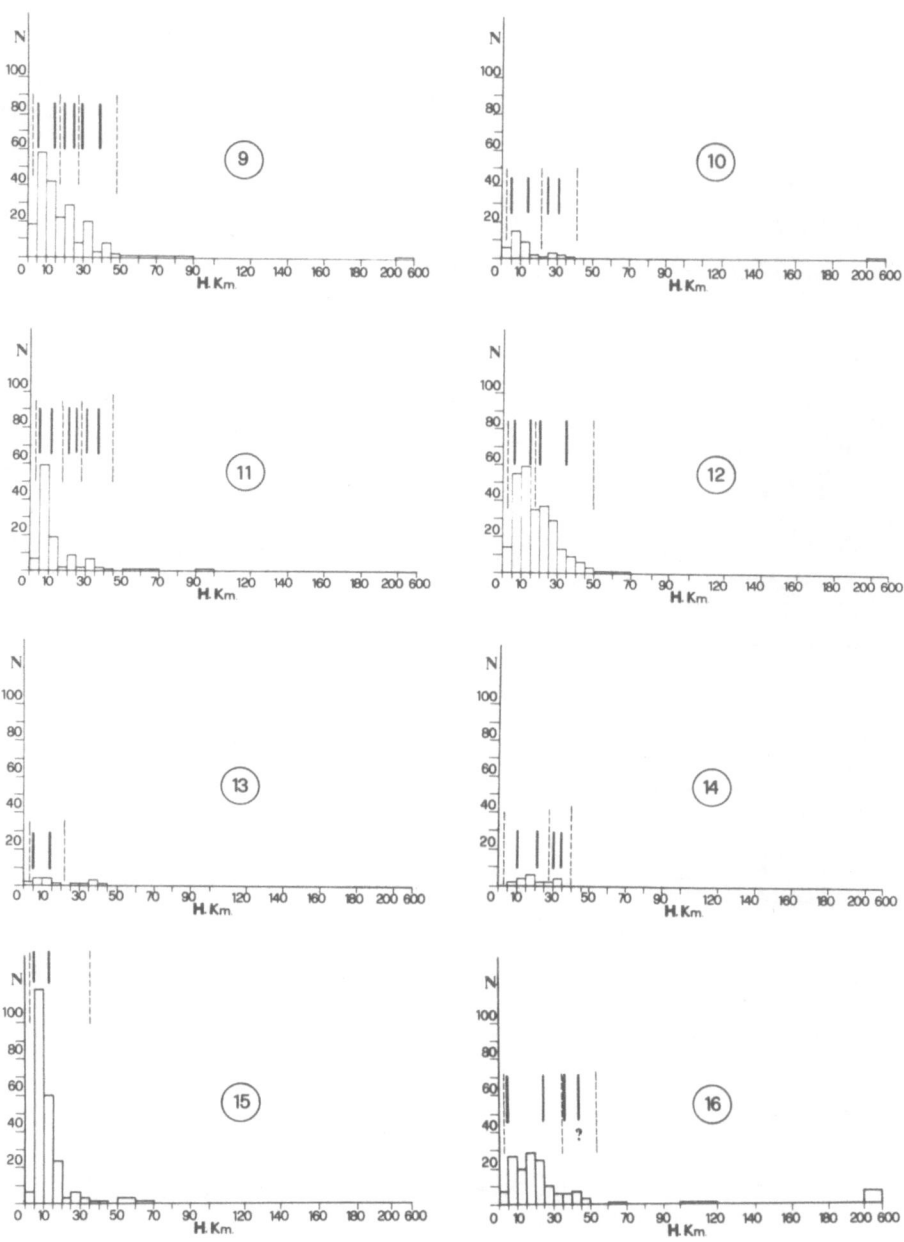

Fig. 8. (continued)

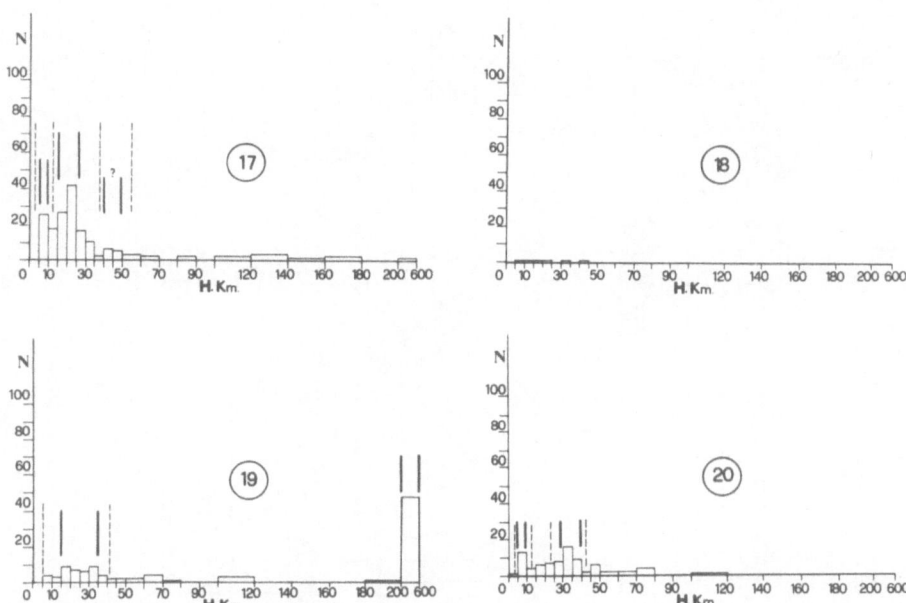

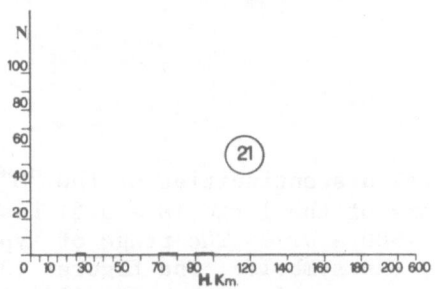

Fig. 8. (continued)

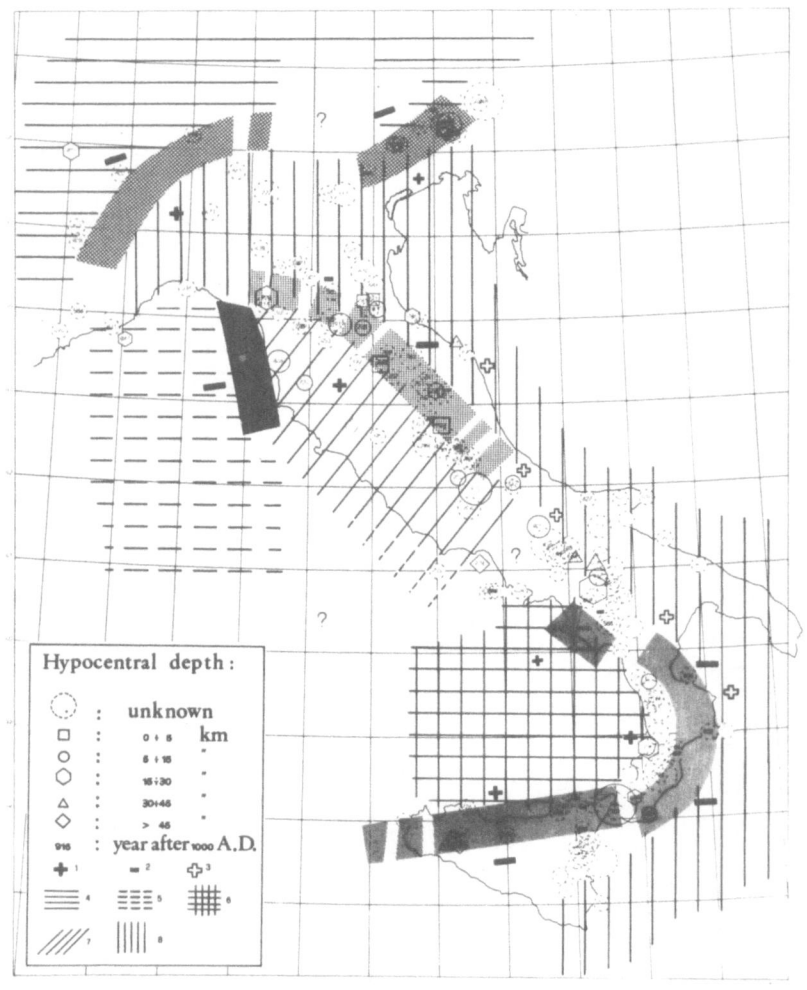

Fig. 9. The major lateral discontinuities of the "M" surface plotted
on the epicenters of the large (M ≥ 5.5) historical earthquakes (1000 - 1980 A.D.). The range of hypocentral depth
is indicated by the symbols. The figures (3 digits) indicate
the year of occurrence after the year 1000 A.D. 1 - lifted
"M"; 2 - depressed (M); 3 - main highs of the aeromagnetic
map probably originated by deep sources; 4 - European domain: Alpine crust; 5 - European domain: Ligurian, Sardo-
Corsican crust; 6 - Tyrrhenian crust; 7 - Intermediate
(Tuscan) crust; 8 - African domain (foredeep and foreland).

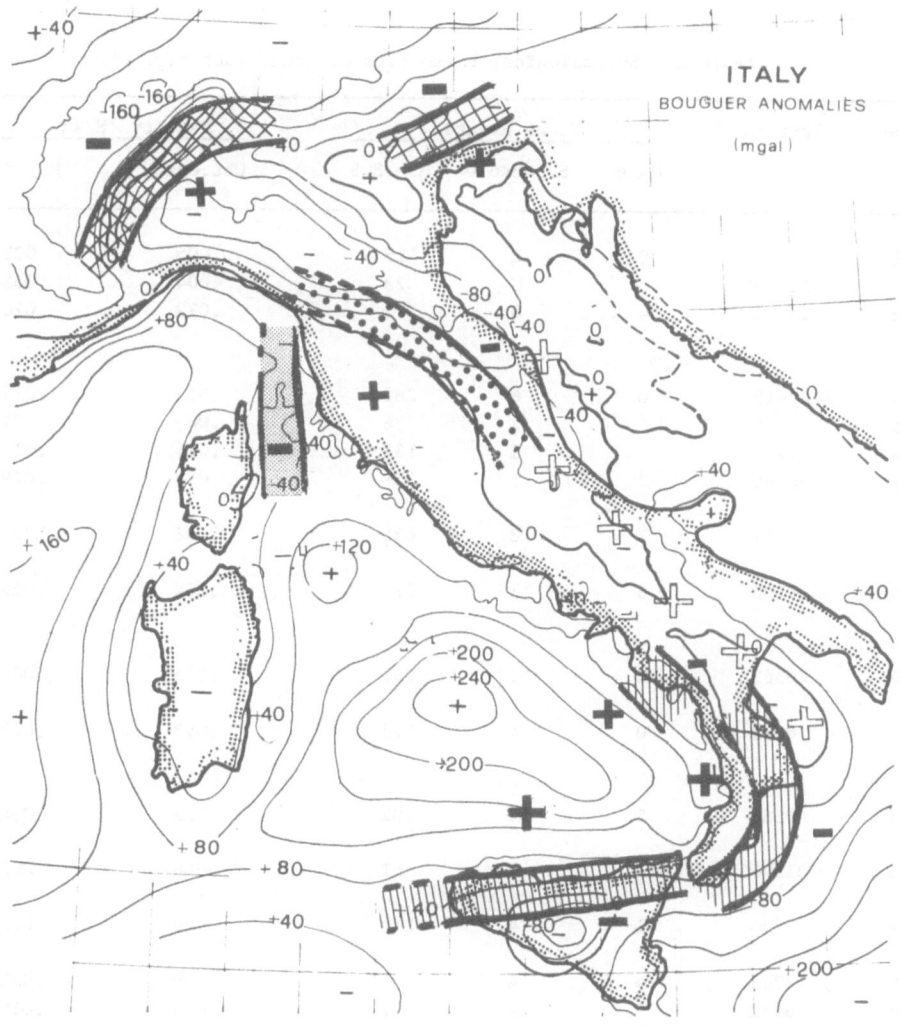

Fig. 10. Lateral discontinuities of the "M" surface plotted on the
simplified map of Bouguer anomalies of reference[17]. White
crosses correspond to aeromagnetic highs.

areas are discarded where the N_{tot} is too small (1, 2, 5, and 21),
an average value of 0.85 is obtained. In Figure 11 the normalized
values of $\underline{a}$ (per km^2) are plotted against $\underline{b}$ values according to
reference[14]. The fitting to the regression straight line is excel-
lent except for the aforesaid zones 1, 2, 21 and for zone 16. The
last one (Southern part of S. Apennines) shows an anomalously low
value of $\underline{b}$ that, considering the seismic activity, seems to indicate

Table 1. Seismological Parametres of Zones (see Fig. 6).

Zone	Surface Km²	N. of Events			Density (N/Km²)	
		M ⩾ 6	5.5<M<6.0	M ⩾ 3.5	Obs.epic.	M ⩾ 3.5
1	7675	0	1	23	.005	.003
2	13893	0	0	23	.004	.002
3	14283	4	8	372	.072	.026
4	23415	0	8	282	.038	.012
5	8872	0	1	36	.009	.004
6	7701	2	1	132	.035	.017
7	38166	2	7	339	.028	.009
8	13867	1	8	436	.086	.031
9	34706	5	11	977	.102	.028
10	26563	0	6	251	.025	.009
11	6478	0	4	178	.093	.027
12	9886	5	11	602	.188	.064
13	31376	1	1	81	.008	.003
14	5750	1	4	115	.050	.020
15	13763	4	4	436	.099	.031
16	18472	9	14	347	.043	.019
17	24664	8	16	851	.114	.034
18	10381	1	3	76	.017	.007
19	27031	2	3	148	.011	.005
20	28280	3	11	661	.052	.024
21	9990	0	0	17	.010	.002

Zone	a	a/km²	b M>3.5	Hypo range(s) km	M_{max} (observ.)	Released energy 1'000 - 1'980 A.D.	
						$\sqrt{Erg}$ tot.	$\sqrt{Erg}/km^2$
1	3.54	+0.452	0.61	7-15	5.8	$.68 \times 10''$	8.86×10^6
2	5.37	+1.237	1.14	10-16	4.7	$.27 \times 10''$	1.94×10^6
3	5.45	+1.275	0.80	10-16 28-36 50-61	6.2	$.85 \times 10^{12}$	5.95×10^7
4	5.26	+0.890	0.79	25-15	5.7	$.62 \times 10^{12}$	2.65×10^7
5	4.54	+0.592	0.80	7-15	5.7	$.83 \times 10''$	9.35×10^6
6	5.13	+1.243	0.81	5-27	6.2	$.25 \times 10^{12}$	3.25×10^7
7	5.96	+1.378	0.93	8-15 30-40	6.1	$.76 \times 10^{12}$	1.99×10^7
8	6.47	+2.328	1.02	5-17 30-44	6.2	$.98 \times 10^{12}$	7.07×10^7
9	6.41	+1.869	0.95	5-14 19-25 28-38	6.2	$.18 \times 10^{13}$	5.19×10^7
10	6.04	+1.615	0.99	5-14 25-31	5.9	$.48 \times 10^{12}$	1.81×10^7
11	5.71	+1.898	0.94	5-11 20-34 30-36	5.7	$.36 \times 10^{12}$	5.56×10^7
12	5.86	+1.865	0.86	6-15 20-35	6.1	$.13 \times 10^{13}$	1.31×10^8
13	5.01	+0.513	0.84	5-14 35-40	6.1	$.21 \times 10^{12}$	$6.69.10^6$
14	5.02	+1.26	0.82	10-21 30-34	6.1	$.25 \times 10^{12}$	$4.35.10^7$
15	5.90	+1.761	0.91	5-13	6.6	$.87 \times 10^{12}$	$6.32.10^7$
16	5.06	+0.793	0.71	5-25 37-45	6.6	$.96 \times 10^{12}$	$5.20.10^7$
17	6.04	+1.648	0.87	15-26 5-9 40-48	6.9	$.19 \times 10^3$	7.70×10^7
18	4.67	+0.653	0.76	?	6.1	$.21 \times 10^{12}$	2.02×10^7
19	4.77	+0.338	0.75	200-600 15-34	6.2	$.31 \times 10^{12}$	1.15×10^7
20	6.26	+1.808	0.96	28-40 5-9	6.6	$.11 \times 10^{13}$	3.89×10^7
21	4.19	+0.190	0.81	?	5.1	$.40 \times 10^{11}$	4.00×10^6

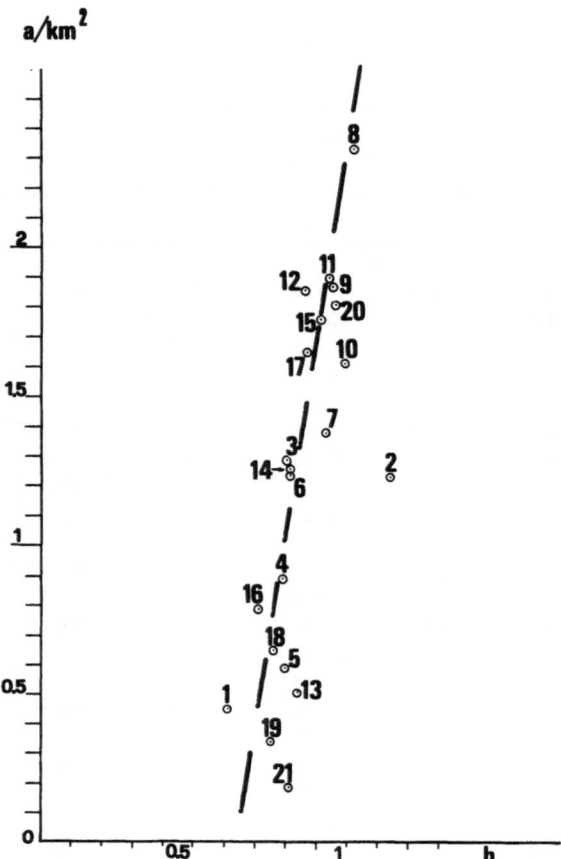

Fig. 11. Normalized values of $\underline{a}$ plotted versus $\underline{b}$ for the zones of Figure 6 (see text for explanation).

a trend towards high magnitudes. For the zones fitting to the straight line the assumption that, for active areas, $\underline{b}$ increases with $\underline{a}$, seems to be confirmed.

- Range of hypocentral depth: in Figure 8 the histograms have been plotted of the number of events for each computed focal depth h for each zone (in steps of 5 km). The most probable range of hypocentral depth has been defined as the one where falls 70% of the total number of events. If the histogram clearly shows more than one peak, the ranges are defined considering 70% of the events occurring between the minima. The upper boundary has been put in any case at h = 2.5 km. It is clear that the definition of depth range(s) is very uncertain, as the existence of sources at different levels also depends on the conditions of observation of the earthquakes.

However, in zones where the number of hypocenters is sufficiently large, some regularities are observed that can be in agreement with the crustal structure. The prevailing range in northern and central Italy is 10-15 km (upper crust) followed by the range 5-10 (mainly sedimentary). Some areas (3, 9) where the activity is high, clearly exhibit a second or a third shorter peak (corresponding generally to the lower crust or to the transition to "M"). This seems to happen especially where a complex crustal structure exists. A sharp change seems to occur along the south Apenninic range. In zone 15 a single range is found at 5-13 km, while in zones 16, 17 and 20 (Irpinia, Calabrian arc and Northern Sicily) the deeper intervals are more active showing a higher seismicity in the lower crust. Finally, in the Aeolian arc the sources are located mainly in the mantle, the most active zone being at depths between 200 and 400 km.

- Energy. The average energy released per km^2 in each zone seems to be a parameter relatively independent from the spatial fluctuations depending on the inhomogeneous accuracy. In Figure 12 the Benioff strain release curves for each zone have been plotted. Individual events with $M \geqslant 6$ are shown by solid segments to give a first idea of the return period. The considered period has been taken as equal for all zones (1301-1971) as well as the time interval (10 years). However, it is clear that in some zones the data are reliable starting only from 1700 A.D. The vertical scales are variable in order to maintain the graphs of the same size. This presentation does not facilitate, sometimes, the comparison between the energy regimes of adjacent zones, that in some cases, is very striking. As an example, in Figure 13 the strain-release curves of North Tuscany and N. Apennines are shown in the same scale for the period 1890-1975 considering a Δt of one year. The behavior of the energy reveals a completely different regime, the release being much faster in N. Apennines than in N. Tuscany where higher magnitudes are generated.

A classification of the zones according to the seismic activity is beyond the scope of the present paper. However, considering only the seismological parameters of Table 2 a scale of severity among the most seismic zones can be established as follows:

1st	17	- Calabro-Peloritan arc
2nd	16	- S. Apennines (Campania)
3rd	12	- Umbrian arc
4th	9	- North Tuscany
5th	20	- North Sicily
6th	3	- South eastern Alps
7th	15	- S. Apennines (Abruzzi)
8th	8	- N. Apennines

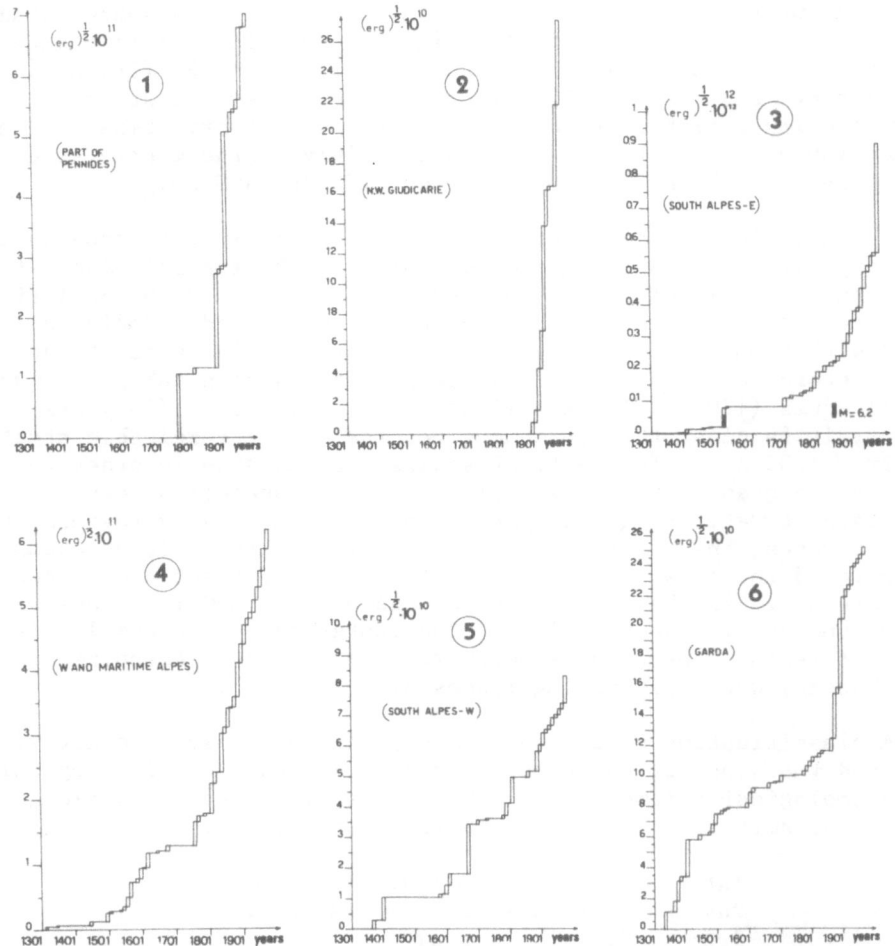

Fig. 12. Strain release curves for the zones of Figure 6
          (ΔT = 10 years). Individual events with M ⩾ 6 have been
          indicated.

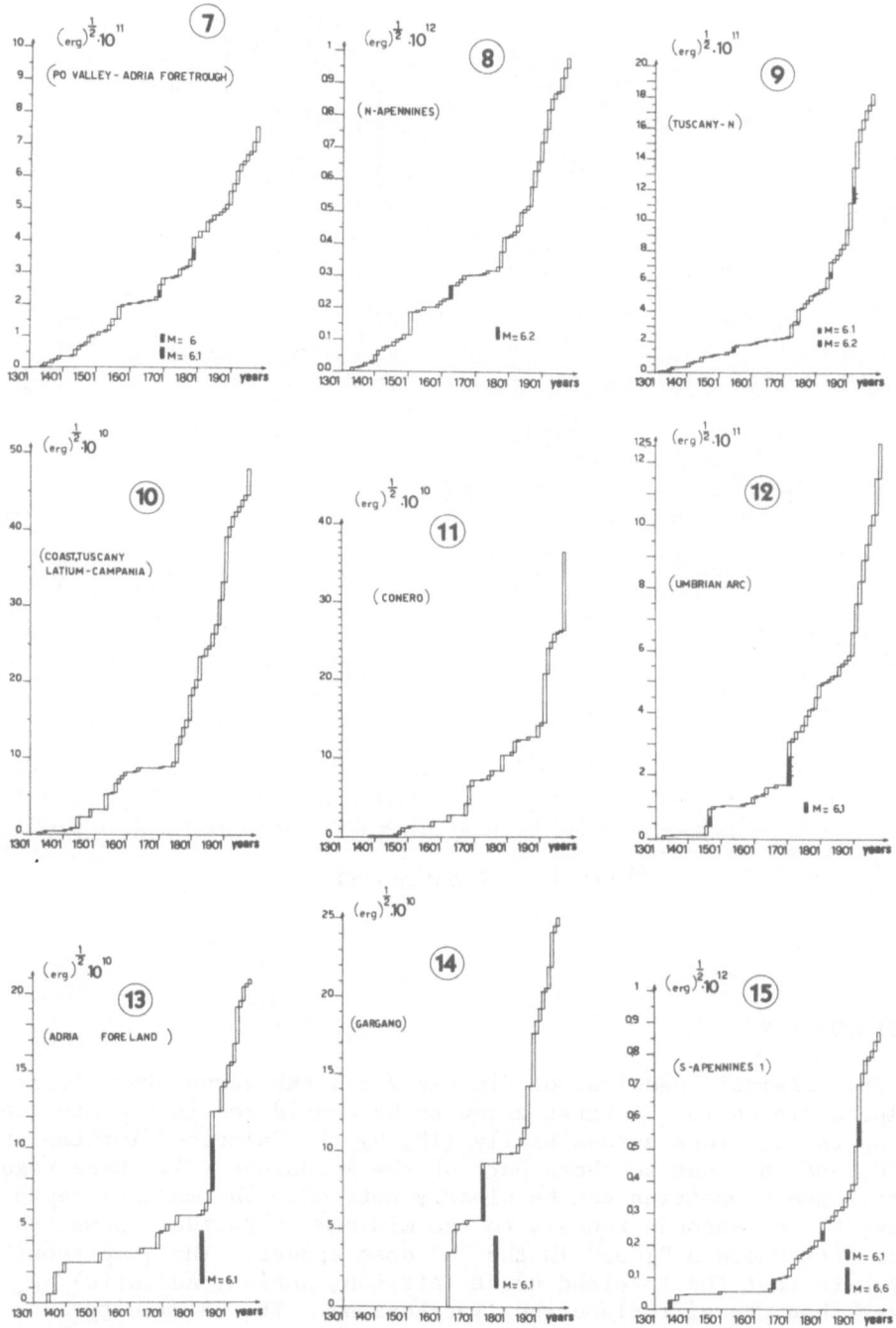

Fig. 12.   (continued).

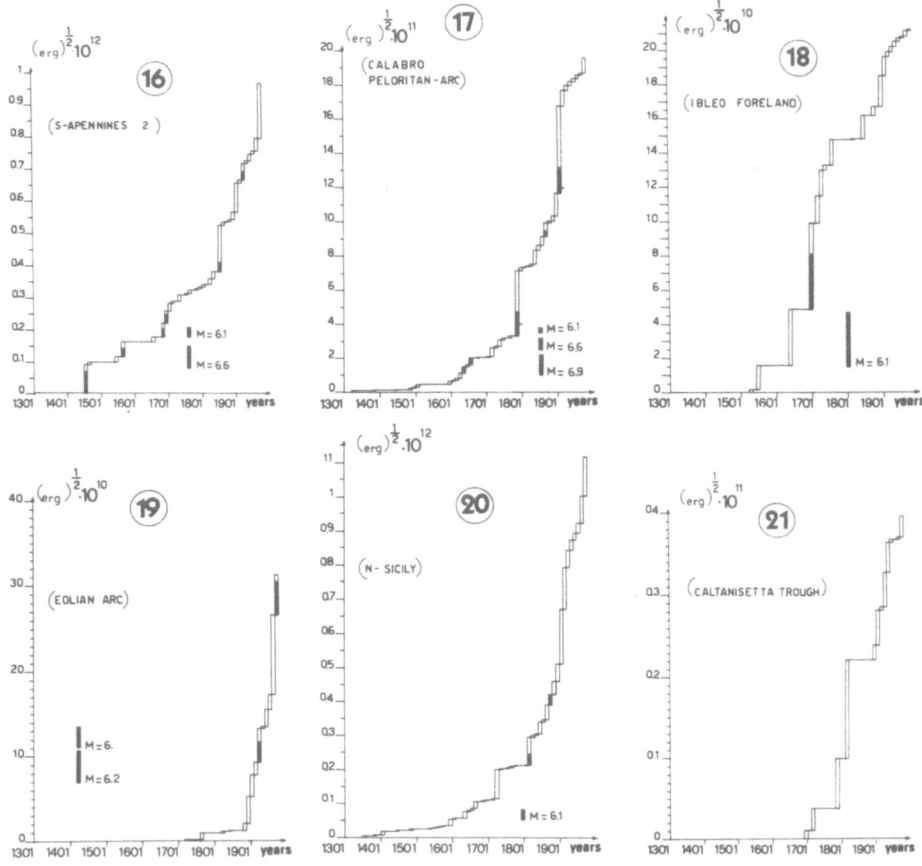

Fig. 12.   (continued).

7.   DISCUSSION

The schematic sections of Figures 7 all run across the Alpine and Apenninic chain. A first group to be considered is the one containing the sections across Sicily (10, 9) the Calabro-Peloritan arc (13, 8) and the most southern part of the Apennines (7). Here regularities and symmetries can be clearly noticed. The maximum depth of the "M" corresponds roughly to the minimum of Bouguer anomalies in the area where a "jump" in the "M" does appear. The jump shows everywhere that the foreland crust (African, Jonian, Adriatic) is thicker than the hinterland one (Tyrrhenian). The seismicity concentrates in the crust or at its bottom in the zone where the "jump" is located.

82

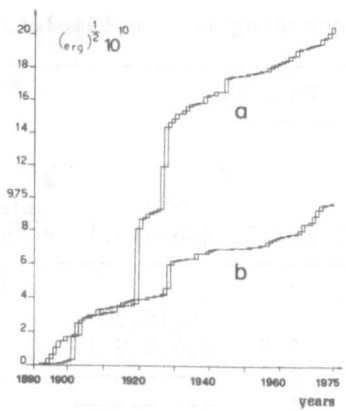

Fig. 13. Strain-release curves (1890–1975) of two adjacent zones
(a – North Tuscany, b – North Apennines). ΔT = 1 year.

The heavy hatched areas represent the depth ranges where the
seismicity is stronger. While in South Apennines (line 7) the major
shocks seem to take place in the upper crust, in the Calabrian arc
as well as in North Sicily, the sources seem situated mainly in the
lower crust. North of zone N. 16, the pattern changes. The SW part
of line 6 (Latina-Pescara) seems to show a thin continental crust;
a small "jump" could be observed at the boundary between zone 15 and
the Adriatic foretrough.

The seismicity, that is very weak along the coast of Latium, is
strong under the Apennines (zone 15) but is shallower than in zone
16. It can be noted that the high seismicity corresponds to the
western slope of the negative Bouguer anomaly towards the Adriatic
foretrough.

Northwards, the next section (5) illustrates the crustal structure between Corsica and Ancona (Conero) through the Elba channel
and Tuscany where the Bouguer anomalies show a flat, slightly positive, pattern. A large "jump" has been found under the Elba channel,
separating a thickening Corsica crust from the thin continental crust
of Tuscany.

Proceeding eastwards, a discontinuity is observed in the "M"
approximately corresponding to a negative gravity gradient under the
city of Perugia. This type of structure is similar to the one found
on line 6.

The thin continental crust of Tuscany has been interpreted[15] as
belonging to the adriatic domain detached from the Adriatic mantle
by a process of antithetic continental collision between the Sardo-
Corsican block and the foreland.

Table 2. Rank of zones according to 8 seismological parameters

Rank	N. of events			Density	M_{max} (observ.)	Released energy $\sqrt{erg/km^2}$	(1) Anomalous b → a	(2) Prevailing depth range→ M_{max}
	M≥6	5.5<M<60	M≥3.5	M≥3.5				
1st	16*	17	9	12	17	12	16	17
2nd	17	16	17	17	15,16,20	17	12	16
3rd	9,12	9,12,20	20	15,8	3,6,8,9,19	8	3	20
4th	3,15	3,4,8	12	9	----------	15	--	--
5th	20	7	8,15	11	----------	3	--	--
6th	6,7,19	10	3	3	----------	11	--	--

(1) b low in respect to a for active areas (see Figure 11).
(2) the zones where the highest magnitudes and deeper (crustal) foci occur rank first.
* zone number.

The seismicity, which is very low everywhere off the Tyrrhenian coast and along the coastal region of Tuscany, Latium and Campania, increases eastwards always being shallow and reaches a maximum in the Umbrian arc. Then, after the relatively aseismic Adriatic fore-trough, the active Gargano strip is met.

The Tuscan zone (N) seems to narrow northwestwards; one of the problems still unsolved is the transition between the Adriatic crust under the Po Valley and the Ligurian sea; in other words, the NW termination of the "Tuscan" crust under the Apennines. The seismicity in Tuscany corresponds mainly to the "grabens" developing NW-SE. In the central part of Tuscany several grabens exist, the more external of which exhibit a higher seismicity (Mugello, Casentino, Val Tiberina) while in the northern part there are only two grabens (Era and Elsa - Garfagnana). While the seismicity in the former is shallow, the prevailing hypocentral depth of the Garfagnana is higher and, in the coastal strip, the Era seems quiescent. The cross section of Figure 14 illustrates in detail the distribution of foci and the available information on crustal structure, up to the top of Mesozoic sediments.

Northwest of the "termination" of the Tuscan zone (probably corresponding to the surface "Levanto-Ottone" line where the ophyolites outcrop), the seismicity falls to very low levels, comparable to the ones of the Po Valley. For this reason, as far as the seismicity is concerned, the transitional zone between the North Apennines and the Maritime and Western Alps have been incorporated into zone 7 (Po valley - Adria foretrough).

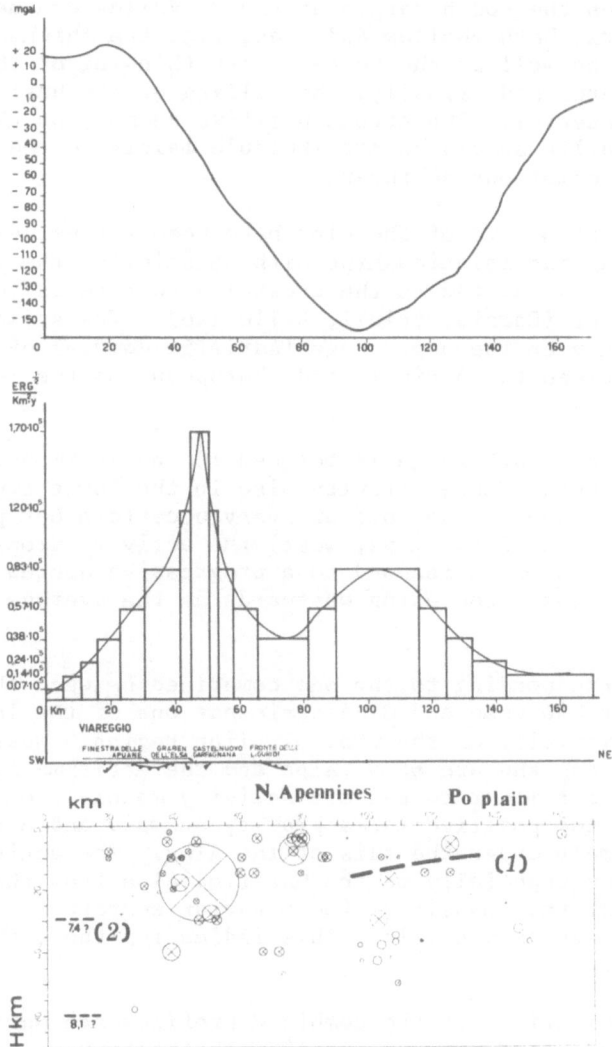

Fig. 14. Cross section through N. Tuscany and N. Apennines, from the Ligurian sea (left) to the Po Valley. Foci falling within 20 km on both sides are projected on the section. From the top: Bouguer anomaly; tectonic flow (after);[20] hypocenters. (1) - Approximate position of the top of mesozoic carbonates. (2) - Velocities of P waves in km/sec.

The profiles 1, 3 show an attempt of correlation between the North Tyrrenian and the South Alps through North Apennines and the Po Valley, where crustal data are not available. The strong negative

Bouguer anomaly on the south margin of the Po Valley can be attributed to several factors, both shallow and deep, i.e. the thickening of the crust northwards as well as the contemporary thinning of the Mesozoic formation southwards and, finally, the filling of the basins with light detrital material. The strong positive anomaly on the North side of the Po valley should be attributable mainly to the outcropping of the Mesozoic formations northwards.

The crustal structure of the Alps have been extensively studied.[13] As regards relationships with seismicity, the area where the foci are clearly related to the crustal structure is the one of the S. Eastern Alps (Carnia, Friuli, Bellunese). The seismicity is concentrated mainly on the upper edge (Adriatic domain) of the crustal discontinuity between the Adriatic and "European" plates (see Figure 4).

The prevailing depth range is between 10 and 16 km but it seems that there is a considerable activity also in the lower crust. In the Garda region (zone 6) the foci are very uncertain but probably shallow, while in zone 5 (S. Alps, west) the activity drops at very low levels. This could correspond to a progressive decrease of rigidity of the upper crust proceeding westwards as the average velocity could indicate.

The area corresponding to the one comprised between the surface tectonic lines of Canavese and Giudicarie has one of the lowest levels of historical seismicity of the whole Italian region. West of the Canavese line, along the arc of W. Alps and the Maritime Alps (zone 4), there is again a moderate seismic activity mainly along the western flank of the positive Ivrea gravity anomaly and also along the negative anomaly under the axis of the Alps. The activity increases southwards especially on the Maritime Alps from the Mercantour-Argentera external massif to the Ligurian western coast.[16] The majority of foci are very shallow, this indicating the stability of the Alpine roots.

The main indications of the combined profiles are partly summarized in the maps of Figures 9 and 10. In the former the position of the sharp lateral discontinuities of the "M" have been plotted on the map of epicenters with $M \geqslant 5.5$. Owing to the uncertainty and also considering the other hypotheses (doubling, flake tectonics, crustmantle "mélange"), the position of the discontinuity has been indicated by a hatched strip 50 km wide. The correlations between the jumps seems clear on the S. Eastern Alps, westwards of line 1, the discontinuity looks shifted to the north and then appears bent along the arc of western Alps. The Tuscan region seems outlined by the sharp discontinuity along the Ligurian coast and the Elba channel (very low seismic activity) and, on the Eastern side, by the slight jumps of Perugia and west of Pescara were the seismicity is high. The continuation of the discontinuity northwestwards is purely speculative. However, considering the behavior of gravity and of seismicity, one could see the crust of Tuscany narrowing and finally disappearing while the Adriatic (Padanian) and Ligurian crust come into direct contact.

The South termination of the intermediate Crust (Tuscan) is not known. Certainly, the picture in the South Tyrrhenian basin and along the S. Apennines looks completely different. A sharp discontinuity exists as in the Elba channel but here the Tyrrhenian crust is much thinner than that of the Adriatic foreland. The same happens along the Calabro Sicilian arc, the foreland Ionian Crust (continental) being thicker than the Adriatic one. The discontinuity in Calabria seems offset eastwards also in respect to North Sicily. In Calabria and in N. Sicily the seismicity clearly concentrates along the Tyrrhenian margin (hinterland) but in the lower crust.

In Figure 10 the same discontinuities are plotted on a simplified map of Bouguer anomalies where the main highs of the aeromagnetic map (given by deep sources) have also been indicated. In the Alps a general agreement exists between the shifting of the crustal discontinuity and the anomalies: however the combined effect of sources at various depths makes difficult the interpretation in the S. Eastern Alps. In the Apenninic range and in the Tyrrhenian the anomalies clearly outline the separation of the two basins and the area occupied by the "Tuscan" crust. As far as the significance of the magnetic highs is concerned it is interesting to compare the latter with the map of the filtered Bouguer anomalies of Figure 15.[17] The pass-band used is $100 < \lambda < 215$ km; therefore the most influencing sources should be, as an average, in the upper crust. The magnetic highs, in a general way, correspond to the gravity lows; this would indicate a rise of the crystalline basement and a thickening of the upper crust eastwards. Except for the Gargano the area east of the line of magnetic highs exhibits a very reduced seismicity (one of the lowest of the Italian region).

CONCLUSIONS

This first attempt is intended as a test of the available data in order to evaluate their suitability to a further statistical analysis. The zoning has been made mainly on the bases of seismological historical data; however, the characters of each area seem also in agreement with the main geological features found by geophysical methods. The present seismic activity is only partially related to the deep crustal structure, the sources being often located in the sedimentary layers, especially the carbonatic ones that sometimes are very thick and of complex tectonics, as in north and central Apennines or along the northern margin of the Po Valley. However, it can be said that the earthquake sources of high magnitude are all related to the crustal discontinuities (S. eastern Alps, N. Tuscany,

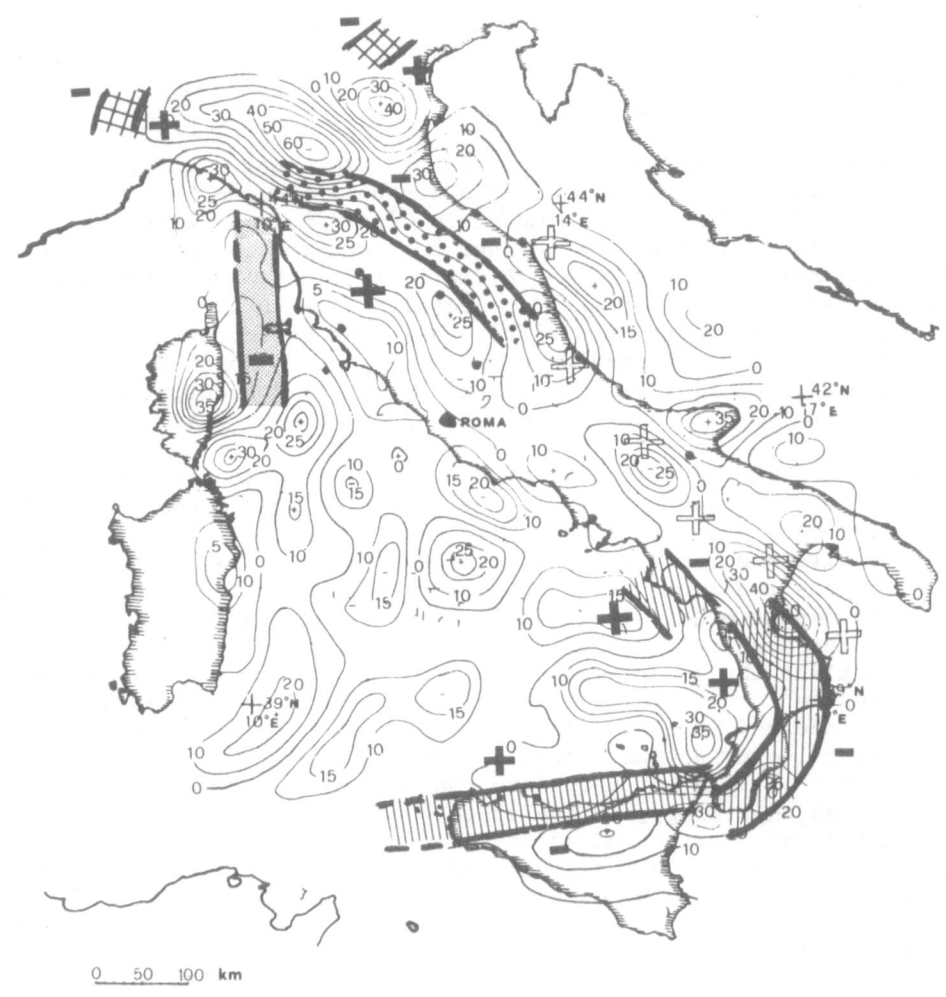

Fig. 15. Same as in Figure 10 plotted on residual Bouguer anomalies.
Pass band of the bidimentional filter 100 < λ < 215 km
(after).[17]   Also the main magnetic highs have been indicated.

S. Apennines, Calabrian arc, N. Sicily). On the contrary, the existence of crustal discontinuities does not correspond to a strong activity in two cases; the first is met along the Tuscan coast and the
Elba channel. Here the seismicity has been probably shifted to the
east margin of the Tuscan area. The second case is the one of the
Ivrea zone and of western Alps were the seismicity is shallow and
the magnitudes low. However, the foci are partly related to the
Ivrea body.

The comparison of seismicity with the geophysical results suggests new topics for a further exploration program, namely:

- the investigation of the seismic "minigap" between the Canavese and the Giudicarie lines in S. western and central Alps where the "M" discontinuity appears shifted;
- the study of the transitional area between Maritime Alps and North Apennines. Here the combination of the available information suggests a continuation of the "Tuscan" crust northwestwards with a possible wedge-shaped termination between the Ligurian (or Corsican) crust and N. Apennines;
- a further investigation of the Tyrrhenian crust south of the Elba channel in order to follow the quiescent boundary between the Sardo-Corsican block and Tuscany, Latium and Campania coastal region. In other words, to define the boundary between the North and South Tyrrhenian basins. The program of the Southern section of the European Geotraverse, foreseen for the Summer of 1983, could answer these questions;
- the present investigation also suggests a major change between the two sections of S. Apennines (Abruzzi and Campania). The boundary between the hinterland (Tyrrhenian) and the foreland (Adriatic foredeep and Apulian "platform") needs further clarification;
- finally, the bending of the Calabrian arc is in agreement with the behavior of seismicity and of crustal discontinuity. Here the Tyrrhenian hinterland and the Adriatic and Ionian foreland seem directly connected. However, the boundary seems shifted eastwards; this fact could agree with one of the proposed models for the arc (Gaudiosi et al., this volume,[18]).

In Sicily the discontinuity seems to continue along the northern strip where the activity is also high. However, further investigation is needed in order to better outline the structure and its possible continuation West and South towards Tunisia and the Sicilian channel. Also here the European geotraverse will give new information. An improvement of the observation of seismicity is also particularly needed in this area.

Acknowledgements

The author thanks N. Tosi who prepared the program and discussed the criteria for the statistical presentation of seismological data, D. Ruspa, who prepared the synthetic cross-sections and A. Tartaglia who took care of the drawings.

REFERENCES

1. P. Giese and C. Morelli, Crustal structure in Italy, Quaderni della Ricerca Scientifica, CNR, 90 (1975).

2. R. Cassinis, R. Franciosi, and S. Scarascia, The structure of the Earth's crust in Italy. A preliminary typology based on seismic data, Boll.Geof.Teor.Appl., XXI, N. 82 (1979).
3. G. Bolis, V. Cappelli and M. Marinelli, Aeromagnetic data on the Italian area, Presented at EAEG, 43rd meeting, Venice, May (1981).
4. M. Pieri and G. Groppi, Subsurface geological structure of the Po Plain, Italy, CNR-AGIP, Progetto Finalizzato Geodinamica, Publ.414 (1981).
5. V. Karnik, Seismicity of the European area, part 1 and 2, Reidel Publ. Ed., Dordrecht, Holland (1971).
6. B. Colombi and S. Scarascia, Sulla interpretazione dei profili sismici crostali. Calcolo diretto della funzione velocità-profondità, Riv.Ital.di Geof., XXII, 3/4 (1973).
7. P. Giese, The determination of the velocity-depth distribution for separated travel-time segments, U.M.P. Committee (1970).
8. N. I. Pavlenkova, Interpretation of refracted waves by the reduced travel time curve method, Izv.Earth Phys., 8 (1973).
9. S. Ballarin, B. Palla, and C. Trombetti, The construction of the gravimetric map of Italy, Publ. of Commissione Geodetica Italiana, Mem. n. 19 (1972).
10. Società Italiana di Mineralogia e Petrologia, Introduction à la géologie générale d'Italie, 26° Congrès Géologique International, Paris (1980).
11. P. Scandone, G. Giunta and V. Liguori, The connection between the Apulia and Sahara continental margins in the southern Apennines and Sicily, Proc. 24th Congr. C.I.E.S.M., Monaco (1974).
12. M. Boccaletti, P. Elter, and G. Guazzone, Plate tectonic models for the development of the western Alps and northern Apennines Nature, Phys.Sc., 234 (1971).
13. P. Giese and C. Prodehl, Main features of crustal structure of the Alps, in: "Explosion Seismology in Central Europe," Springer-Verlag, Berlin-Heidelberg, New York (1976).
14. K. Kaila and H. Narian, A new approach for preparation of quantitative seismicity maps as applied to Alpide Belt-Sunda arc and adjoining areas, Bull.Seism.Soc. of America, 61:5 (1971).
15. K. Reutter, A trench forearc model for the northern Apennines, in: "Sedimentary basins of Mediterranean margins," Tecno-print, Bologna (1980).
16. G. Capponi, C. Eva, and F. Merlanti, Some considerations on seismotectonics of the western Alps, Boll.Geof.Teor.Appl., XXII (1980).
17. G. Corrado and A. Rapolla, The gravity field of Italy: analysis of its spectral composition and delineation of a tridimensional crustal model for Central-Southern Italy, Boll.Geof. Teor.Appl., XXIII, 89 (1981).
18. G. Gaudiosi, G. Luongo, and G. P. Ricciardi, A bending model for the Calabrian arc, in: this volume.

19. G. F. Panza and S. Mueller, The plate boundary between Eurasia and Africa in the Alpine Area, Mem.Sc.Geol., 33 (1978).
20. M. Cattaneo, C. Eva, and F. Merlanti, Seismicity of Northern Italy: a statistical approach, Boll.Geof.Teor.Appl., XXII 89 (1981).
21. C. Morelli, P. Giese, M. T. Carrozzo, B. Colombi, I. Guerra, A. Hirn, H. Letz, R. Nicolich, C. Prodehl, C. Reichert, P. Rower, M. Sapin, S. Scarascia, and P. Wigger, Crustal and Upper Mantle structure of the northern Apennines, the Ligurian sea, and Corsica, derived from seismic and gravimetric data, Boll.Geof.Teor.Appl., 75-76 (1977).

STRUCTURE SEISMOLOGY

I. P. Kosminskaya

Institute of Physics of the Earth Acad. Sci. USSR
B. Gruzinskaya 10
D-242, Moscow, USSR

I. MODIFICATIONS OF THE STRUCTURE SEISMOLOGY

Modern seismology has two main parts: foci seismology and structure seismology. The first part is connected with the study of earthquakes as physical phenomena and their consequences: seismicity and seismic risk. The second is connected with the Earth's structure study using the seismic waves generated by earthquakes and artificial sources.

1.1. Considering the type of wave generating sources structure seismology can be divided into two branches: earthquake waves seismology (ESW) and controlled sources seismology (CSS). Each of these two branches includes some modifications depending on the type of the waves used, subjects of study and systems of observation (Figure 1). The seismic prospecting (SP) part of controlled sources seismology has the highest resolution. SP now uses the common deep point (CDP) operation with a number of one-channel recorders at 1 km profile reaching 10^3 (Table 1). The world wide teleseismic net – part of earthquake wave seismology – with distances of thousands km between stations has the lowest resolution but depth studies by this net are unlimited.[1]

1.2. The deep seismic sounding (DSS) method belongs also to the CSS branch of SS and its resolution in principle is less than in seismic prospecting but higher than in EWS modifications.[2] The main subject of DSS is the study of the crust and upper mantle. DSS uses body P, S, SP and PS waves and a system of observation very close to the system used in the seismic prospecting, including CDP. The main DSS systems of observation are: continuous, semi-continuous and point profiling, and differential sounding.[3] Continuous profiling employs

93

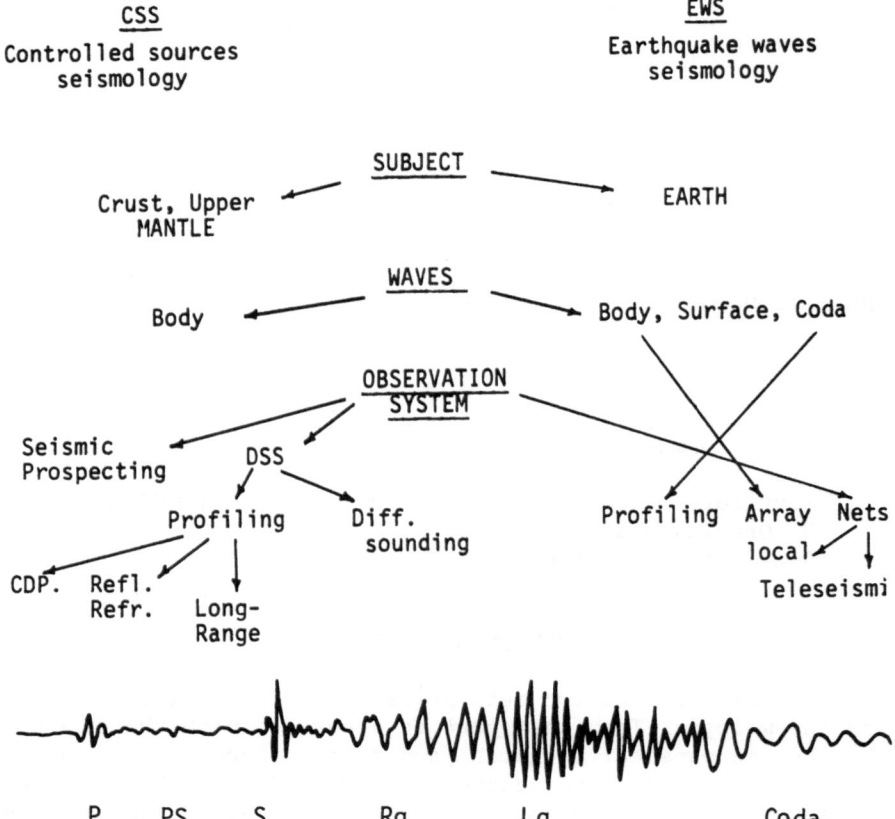

Fig. 1. Modifications of the structure seismology. Below: an earthquake record.

explosions and vibroseis wave generators and densely distributed one- or three-components recorders of refracted and reflected (sub and over-critical) waves. Continuous profiling is common for the detailed study of the upper part of the crust. Point profiling is used at sea[4] and during long shooting (LRP) on land.[5-8] In some cases (as will be shown in the second part) the resolution of LRP is close to that of the EWS observation of the seismological profiling and array (Figure 1). The interpretation of data obtained by the profile system of observation deals with the reversal and overlapping travel-

Table 1. Resolving power of SS modifications (see Figure 1)

METHOD PARAMETERS	SP	DSS	EWS
1	2	3	4
F cps	$10\text{-}10^2$	$1\text{-}10$	$10\text{-}10^{-3}$
= F.V km	$6.10^{-1}\text{-}6.10^{-3}$	$6\text{-}6.10^{-1}$	$6\text{-}6.10^{3}*$
V = 6 km/s			
h/	10^{-2}	$1\text{-}10$	$1\text{-}10^{-2}$
N /km	10^3	$10\text{-}10^2$	10^{-1}
F.N cps km^{-1}	10^5	$10^{2\text{-}3}$	10

N - number of one channel records on 1 km profile.
*surface waves.

times curves. The more dense the system of T-T curves the more reliable are the seismic cross sections.

Differential sounding uses numbers of separate elementary observations: one source and small array (⁓ 1 km). The distance between the source and array is optimal for the recordings of the main
observed waves: for example, 150-200 km for common recording of reflected and refracted M waves. This modification was developed in
the sixties for seismic observations in complicated site conditions
of the East part of the USSR such as the Taiga area. This modification operates with a special type of travel-time field and permits
the determination of the depth of the main boundaries and layer velocities in the crust.[3]

Routine DSS observations are usually used for the crustal study
to the depth of M discontinuity. The optimal length of travel-time
curves is about 300-400 km for continent and 100-150 km for ocean.
For the detailed mantle study long range profiles are used.

1.3. In the USSR the DSS method was developed towards the end of the
fifties. A lot of DSS profiles crossed the territory of the USSR
(Figure 2). These profiles transected different types of geological
structures: shields, platforms, downwarps with thick sediments etc.
Many DSS profiles were conducted in the marginal zone of the Eurasia
continent and in the oceans (Pacific, Atlantic, Indian), inland seas
(Black sea, Caspian sea, Mediterranean sea), in marginal seas (Bearing
sea), and in some areas of Antarctic.

1.4. A lot of DSS data obtained during more than 30 years needed of
course systematization and generalization. The first step in this
direction was made at the end of the sixties. The first map of the

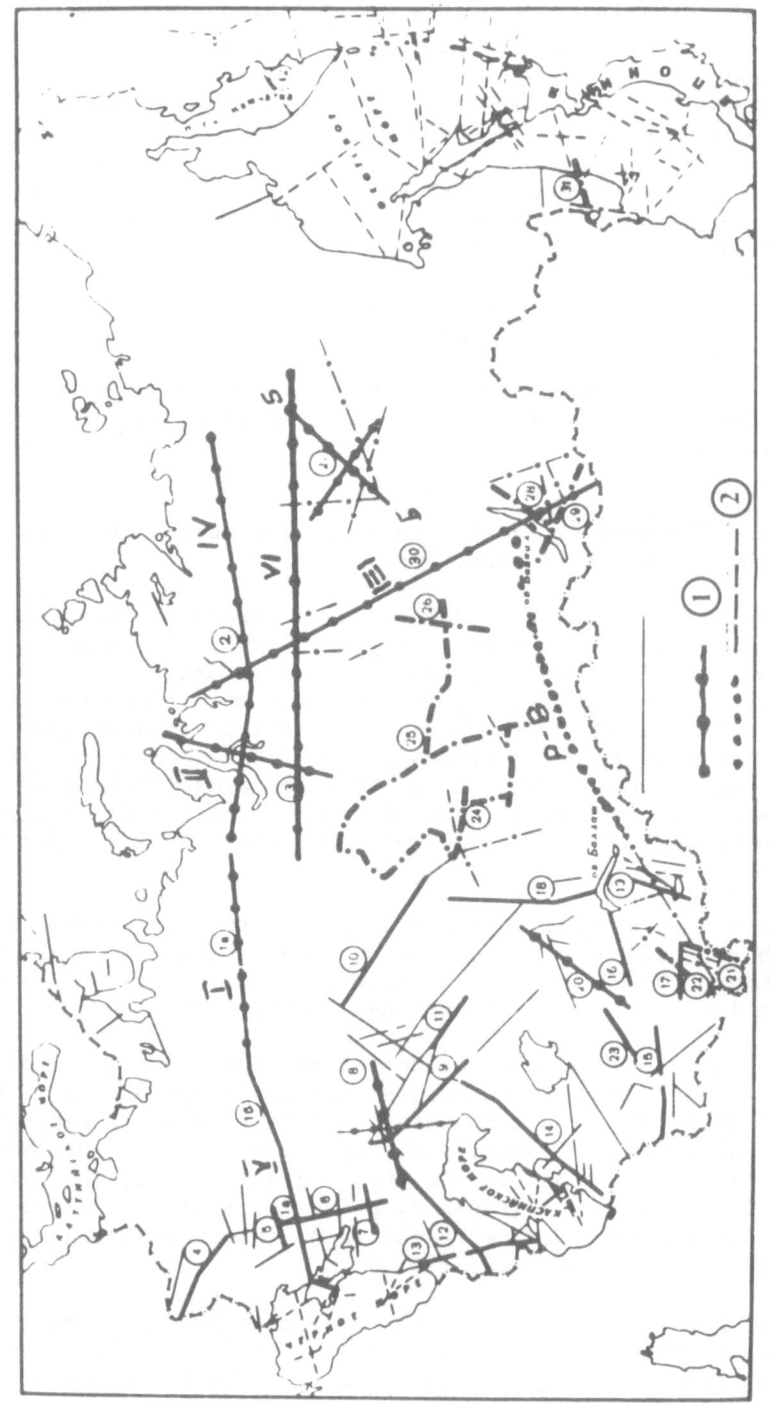

Fig. 2. DSS profiles on the territory of the USSR. 1 – Long range DSS profiles.
2 – Earthquake profiles.

96

relief of M discontinuity for the territory of the USSR on the bases of DSS and gravimetrical data was compiled. New ideas about the wave nature and in the theory of wave propagation led to the second step of compilation of DSS data. Some of the old DSS profiles were reinterpreted and together with the new data presented in the book: "Seismic models of lithosphere for major geostructures on the territory of the USSR".[3] In this book much attention was paid to wave analyses and relations between wavefield and velocity models of different types of geological structures. The problem of the nonuniqueness of the seismic reverse problem for the various kinds of systems of travel time curves were also discussed.

1.5. Using the DSS data compilation the main features of the crust for major continental structures were formulated (Figure 3). Since the shields and platforms represent about 40% of the territory of the continents it is natural to consider the generalized velocity model for these structures as a typical model for the continental crust. This model is 40 km thick and includes at least three layers (not two as was common before). All three layers have about equal thickness (10-14 km) but a different range of velocities: 5.8 - 6.2; 6.4 - 6.7; 6.8 - 7.6 km/s and for the M discontinuity 7.8 - 8.2 km/s. The seismic characteristic of the first layer is very close to the former "granitic" layer. The second and the third layers which in some areas were combined in the "basaltic" layer have different properties. The second layer has unstable parameters: varying and small velocity gradients that sometimes include LVL. The parameters of the third layer are not well known. Its thickness and layer velocity were determined predominantly by near and overcritical reflected waves. Only during the last few years have refracted waves connected with the third layer been observed as the first arrivals (Baltic Shield's profiles: Fennolora-Finlaps 1979; Sveka, 1981; Baltic, 1982).

1.6. In Figure 3 generalized models for different structures of continent and oceans are shown. The orogenic zone (Figure 3a) has a thicker crust than the typical continental one (Figure 3c) and other relations between the three main layers. In some orogenic zone the third layer is the thickest one (Tien-Shan). Some of geologists connected this layer with the low velocity mantle. The thickness of the crust of the Pamir orogenic area reaches 70 km. One may suppose that it is a collision zone, but we have no seismic data on the overlapping process besides the comparatively low average velocity ($\sim$ 6.4 km/s) to the M discontinuity. On the Eurasia continent some regularities between the crustal velocity model of the platform and orogenic zones and their position to marginal area are observed. The structures inside the continent (the Russian platform, the Caucasus and oth.) have a thicker crust than the marginal structures such as the west European platform and the Primorye ridge.

1.7. The new DSS data obtained in the oceans proved the distribution of velocities in the deep oceanic basins to be typical for oceanic

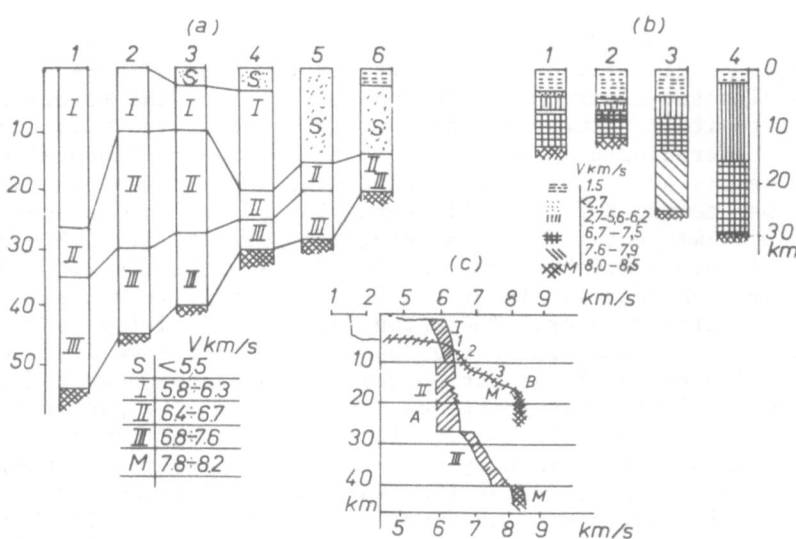

Fig. 3.   Velocity models of continental and oceanic structures.
a) continental: 1 – orogenic zone; 2 – shield, 3 – old
platform, 4 – young platform (West European), 5 – deep
continental down warp, 6 – Insland seas.   b) oceanic:
1 – marginal sea; 2 – oceanic basin, 3 – passive rise
(Shatsky rise); 4 – microcontinent (Lord-How plateau);
c) typical velocity models for continental (A) and
oceanic (B) crust.   I, II, III correspond to the
number of layers at a),b).

crust. The new detailed model based on the generalized data (Figure 3b,c) is very similar to the well known Raitt's model presented in 1956. The main difference of the former and new models is the presence of a high velocity layer (7 - 7.6 km/s) in the lower part of the crust. The thickness of this layer varies from 1 to 3 km and in cases when this layer has a thickness of less than 1 km it can be missed if using only the first arrivals of the refracted waves for interpretation. In the new typical oceanic model the second layer is divided into three parts: 2A, 2B, 2C with velocity 3 - 6 km/s. In some areas layer 2C has a very close characteristic to the 3A layer and the boundary between the second and the third layers in many cases is a good refractor but not a good reflector. The typisation of the model for other oceanic structures is under discussion. It needs new and more detailed data comparable with those obtained at the Eurasia and the North American continents.

Conclusion

Modern structure seismology uses several arrangements to study the different scale of vertical and lateral heterogeneity of the earth's crust and upper mantle (Table 2). New data obtained at the continents and oceans testified the correlation between the crustal velocity models and geological structures, their age and tectonic history. In some cases this correlation is not clear and seismic properties of the crust are similar for different tectonic zones. Perhaps in this case the connection between physical parameters of the crust and geological data is more complicated and needs more detailed study. The new DSS data provides a strong support to the division of all of the seismic models into two parts: continental and oceanic. The structures with continental type crust occupy almost all the continental surface. Only downwarps and rift zone can be considered as areas with intermediate type of crust. The oceanic type of crust represents only the deep basin (about 40% of the oceanic area). The classification of seismic models of the crust of the other oceanic structures such as rift zones, passive rises, microcontinents and Islands needs new more detailed observations. Some years ago we proposed[9] to divide the models of the intermediate types of the crust into two groups concerning the thickness and velocity values: subcontinental and suboceanic. Now this terminology should be improved with addition and more accurate determination of the range of thickness and especially the range of values of velocities for consolidated crust.

II. REVIEW ON LONG-RANGE SEISMIC PROFILES IN THE USSR

2.1. Long Range Profile (LRP) belongs to the DSS method. Field operation on the long range profile includes reversed and overlapping travel time curve (T-T) systems with distances 100 - 200 km between

Table 2. Scale of Heterogeneity of the Crust and Upper Mantle

	LAYERING V(Z)	BLOCKING V(Z,L/X,Y/)
1.	GLOBAL: Crust, uppermost mantle... ⌣50-100-500 km	Continents, oceans, transition zones... ⌣1000 km
2.	MACRO: Layers in the crust and mantle... ⌣5-20 km	Structures inside continent and oceans... ⌣100 km
3.	MESO: Layers inside main layer... ⌣3-10 km	Structures inside platform, orogenic zone... ⌣10 km
4.	MICRO: Fine structures of boundaries... ⌣0.1 - 3 km	Fault zone, micro-tectonic structures... ⌣1 km

shot points (SP) for the crust and 500 - 3000 km for the upper mantle. More than 100 three-component recorders are distributed along the profiles at intervals of 10 - 15 km. The frequency band is 1 - 20 cps. Owing to three-component equipment P, PS and S waves are observed and used for interpretation. The position of LRP on the territory of the USSR is shown in Figure 2. The LRP cross the Russian Platform, young West Siberia and the ancient Siberian platform. Many publications in Russian and English are used in this review but of course only the main results will be discussed.

2.2. Figure 4 shows the generalized record section for the upper mantle P waves at distances from 100 to 3000 km. Space interval between the neighboring traces is about 100 km. Numbers mean the depth. The picture below corresponds to dashed range of distances. Solid lines are reflected and refracted travel time curves calculated for models fitting the observed data. CP-critical points (solid arrows). Records are improved by adaptive filtration. Figure 5 shows S^M waves. There are two shear S^M_X and S^M_Y components, arrival times for which are different. It may be accounted for by different velocities for the waves propagating along and across the profile. One of the most possible explanations is anisotropy of the crust on the whole. Next Figure 6 shows the reversed T-T curves for the given one dimensional model. The main feature of these T-T curves are reflected waves from the upper and lower boundaries of the LVL at a depth of about 200 - 250 km (solid arrows).

Figure 7 presents a cross-section constructed for the eastern part of profile IV (see map Figure 2). This cross-section is typi-

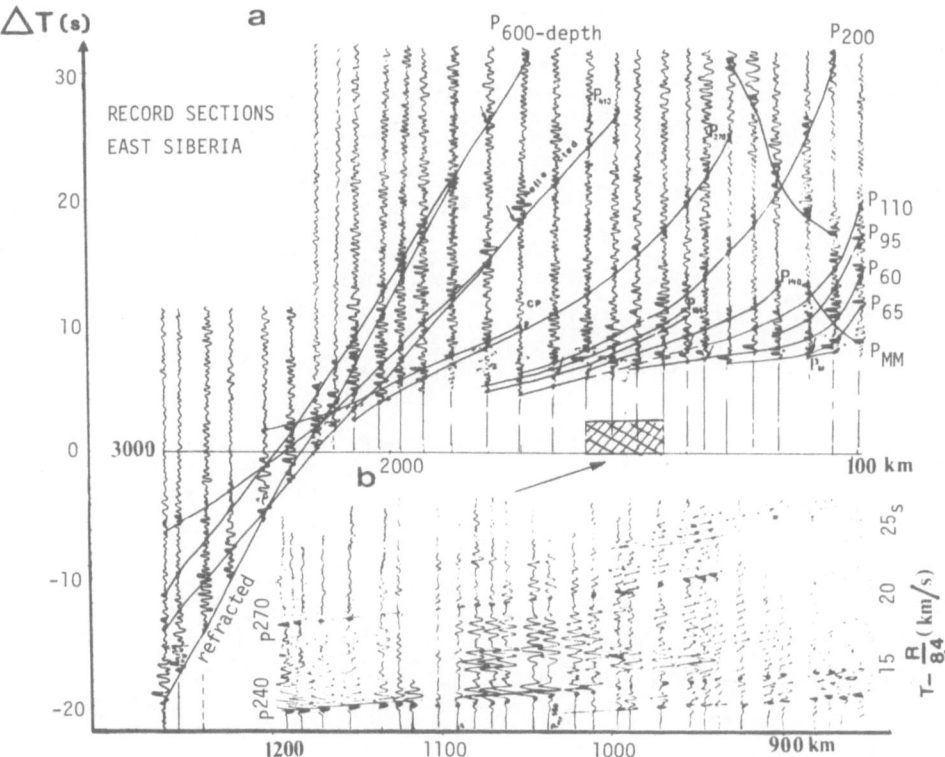

Fig. 4. An example of LRP record section. Below – part of the
section in detail for the distance ⁓ 1000 km (see shadow
zone). Reduction velocity for a) 8.2 km/s, b) 8.4 km/s.
Numbers on the wave symbols (P_{600}....) indicate the depth
of reflectors and refractors (see Figure 10 and 11).

cal of LRP: We can see many unstable discontinuities in the crust
determined by P and PS waves. PS waves permit the determination of
first order discontinuities with a drop or rise of velocities. Reflected and refracted P waves are also used.

The most interesting are velocities' isolines for the uppermost
mantle. There is LVL in which the velocity is comparatively higher
than normal in the mantle.

Figure 8 shows the set of T-T curves observed at some LRP. You
can see that they are quite different from Herrin T-T. Two upper
profiles (I and II) belong to the Russian platform. These are the
largest times which means lowest velocities. On the contrary for
East Siberia small times are observed, that is high velocity. The
difference exists at the distances up to about 2000 km and at larger

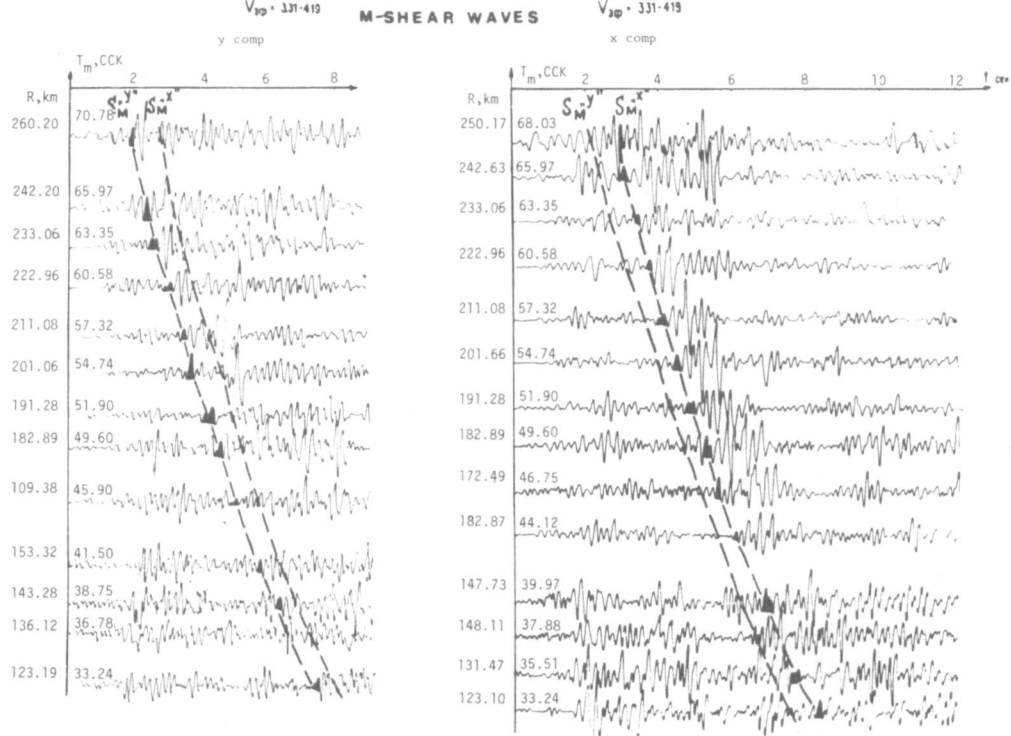

Fig. 5. Examples of S wave record sections. Note the difference between arrival times of S_x and S_y waves which can be accounted for anisotropy.[8]

distances T-T come close to each other and to Herrins T-T. It can be accounted for by homogeneity of the lower part of the upper mantle.

Figure 9 shows the amplitudes for two LRP in East Siberia. In the upper part amplitudes of earthquake waves in the Kazakhstan Baikal area are shown. LRP amplitudes were used for determination of the Q factor.

2.3. Let us consider now one dimensional model for the upper mantle. In Figure 10 models for the lower lithosphere are shown. The first model corresponds to the Northern part of the Russian platform. It is very simple. The next are more complicated. It may depend a little on the quality of data. We draw your attention to the next three models which were obtained with the data of the same quality and very similar T-T systems. At depth LVL and HVL are observed. At longitudinal profile, velocity is low, at latitudinal - high; but in both cases they are higher than in the normal mantle. Authors of these models - Egorkin et al., believe there is evidence of anisotropy phenomena in the low lithosphere.

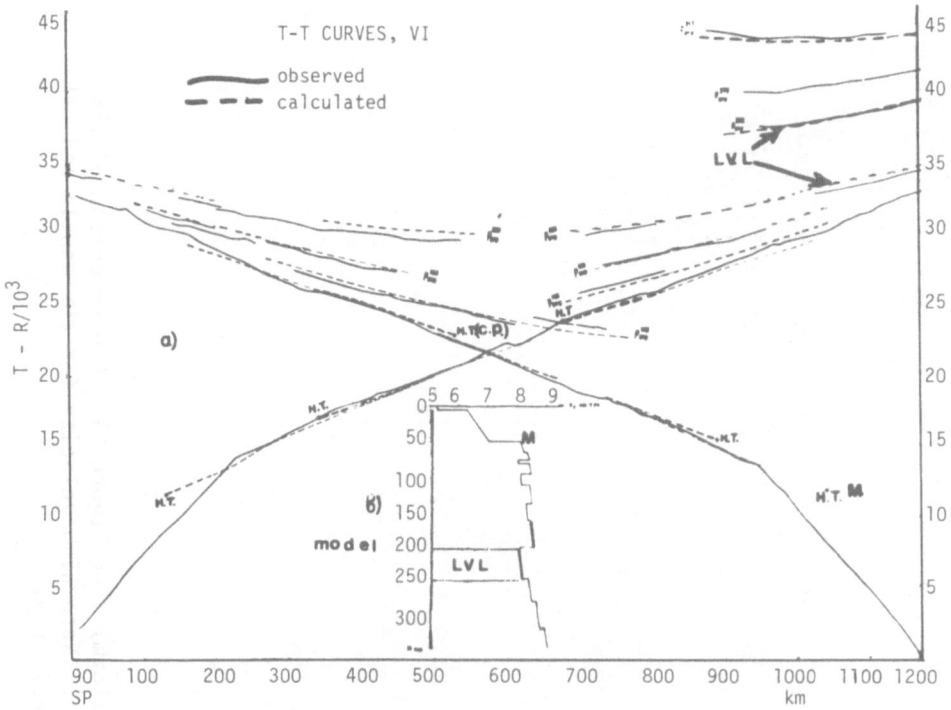

Fig. 6. Example of reversed travel time curves system and one dimensional velocity model including LVL. Travel-time curves
for this model (dashed line) fit with the observed reflected
and refracted waves. WT - critical points.

Figure 11 shows generalized one dimensional models for the upper
mantle to a depth of more than 600 km. There are the same two profiles (III long. and IV lat.) which are discussed above.

LVL at a depth of 200 - 250 km and sharp boundaries at 400 and
600 km are of main interest. In comparison to the LVL in the uppermost part of the mantle this LVL has really low velocity, close to
8 km/s, while the average velocity at this depth is about 8.2 - 8.4
km/s. To the right the Q factor is shown for the same profile. We
can see an increase of Q from 200 to 500 at a depth of 100 - 150 km
and then a decrease to 200 at a depth of 300 km and again a rise to
400 at a depth of 400 km. Curve 3 corresponds to Anderson's data
(1978). The general trend for both curves is very similar but differences in values may depend on accuracy of data and peculiarities
of regional mantle structure.

The final two figures show EWS data obtained in a different kind
of region - Pamir-Baikal orogenic zone. Figure 12 shows a T-T system

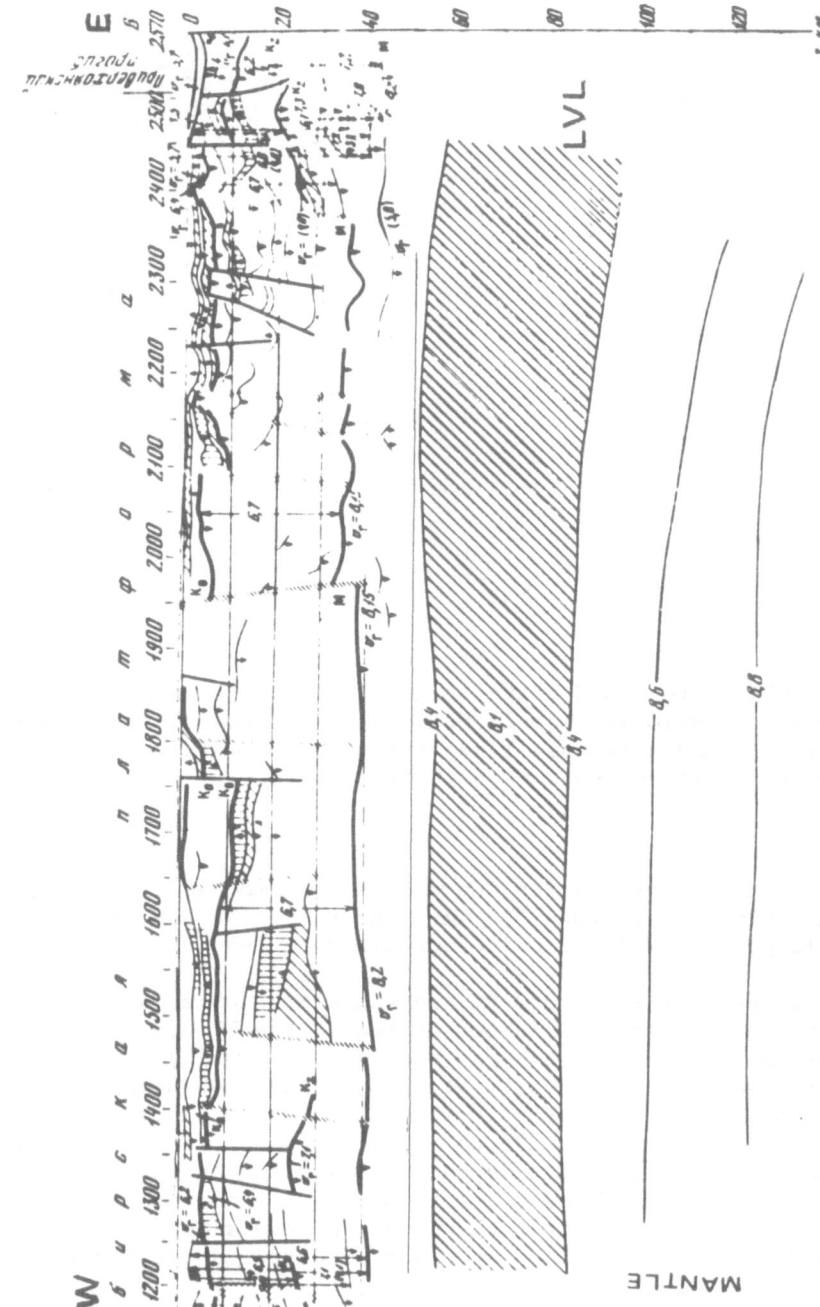

Fig. 7. East part of the LRP VI (Siberian platform) (see Figure 2). Cross-section of the crust and uppermost part of the mantle. Note high velocity for the mantle and LVL with velocity equal to normal mantle velocity.

104

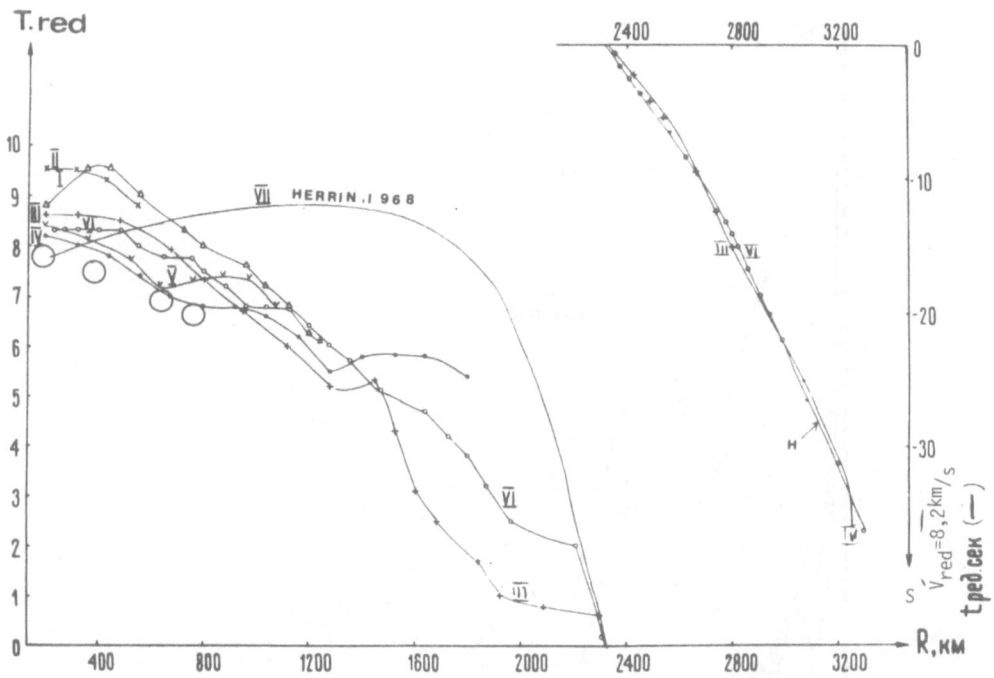

Fig. 8. Travel-time curves for different long-range profiles.
Numbers of curves correspond to the number at Figure 2.
Circles-arrival times at the Fennolora profile 1979, Baltic
Shield. V_{red} = 8.2 km/sec.

of observation. Numbers correspond to the earthquake epicenters.
Figure 13 presents one-dimensional models fitting the observed T-T
curves.

In Figure 11 we compare curves 4 and 5 with LRP data. We observe in general a good agreement between LRP and EWS data. Differences can be explained through differences of the regions and
accuracy. Important is the stable position of the main discontinuities at 400 and 600 km and the value of velocities at these depths.

From the point of view of methodics it is important that there
is a clear possibility to combine both types of data and use them
for the upper mantle study.

Conclusions

Improved technique of seismic operation was used on the LRP,
that increased the depth and detail of the upper mantle study and

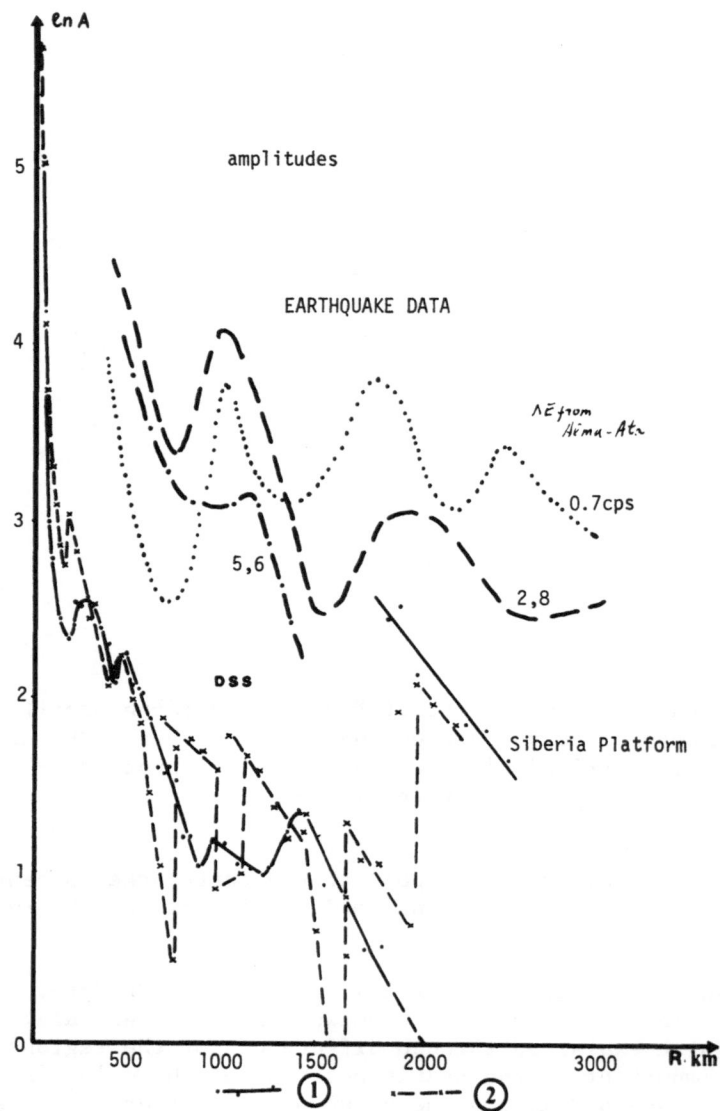

Fig. 9. Amplitude-distance curves for different frequency ranges: earthquake (1) and DSS-LRP data (2). Note similar features of curves for high and low frequencies at the distance about 1000 km and increasing of amplitudes of high frequency at the distance 2000 km.

enabled penetration to more than 600 km depth with the length of T-T more than 3000 km. Sub- and over-critical reflected P and S waves from upper mantle discontinuities were observed and used for

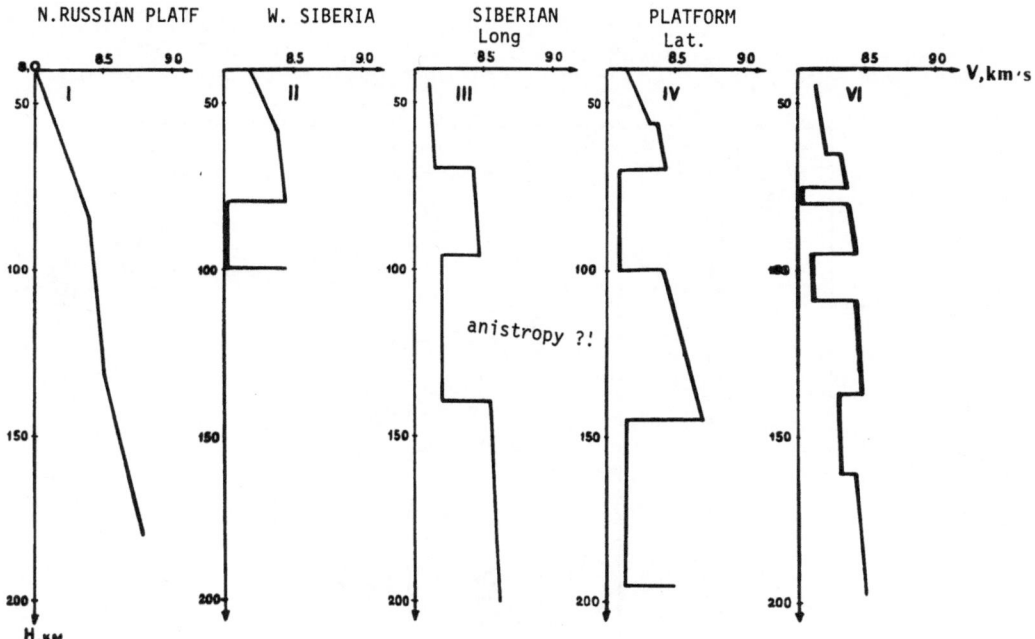

Fig. 10.   One dimensional velocity models for the upper-most part
of the mantle. Note the low and high velocity layers at
longitudinal and latitudinal profiles on the Siberian
platform. The difference can be considered as anisotropy
or more possible the influence of the lateral hererogeneity.

interpretation. A new type of upper mantle model was constructed.
It includes high and LVL with thickness of some dozens of km and
sharp boundaries.

On the ancient Siberia platform different velocities were observed at longitudinal and latitudinal profiles which can be accounted
for by the anisotropy phenomena. At a depth of 200 - 250 km LVL is
observed which may be considered as Gutenberg's asthenosphere.

Many common features in CSS and EWS models are observed. It is
a good reason to combine these two methods for the upper mantle
study. This big success in the upper mantle study enables us to
come nearer to solving the global problem of the mantle roots of
continents and oceans as a whole and to approach an understanding of
what types of upper mantle create different types of continental and
oceanic crust.

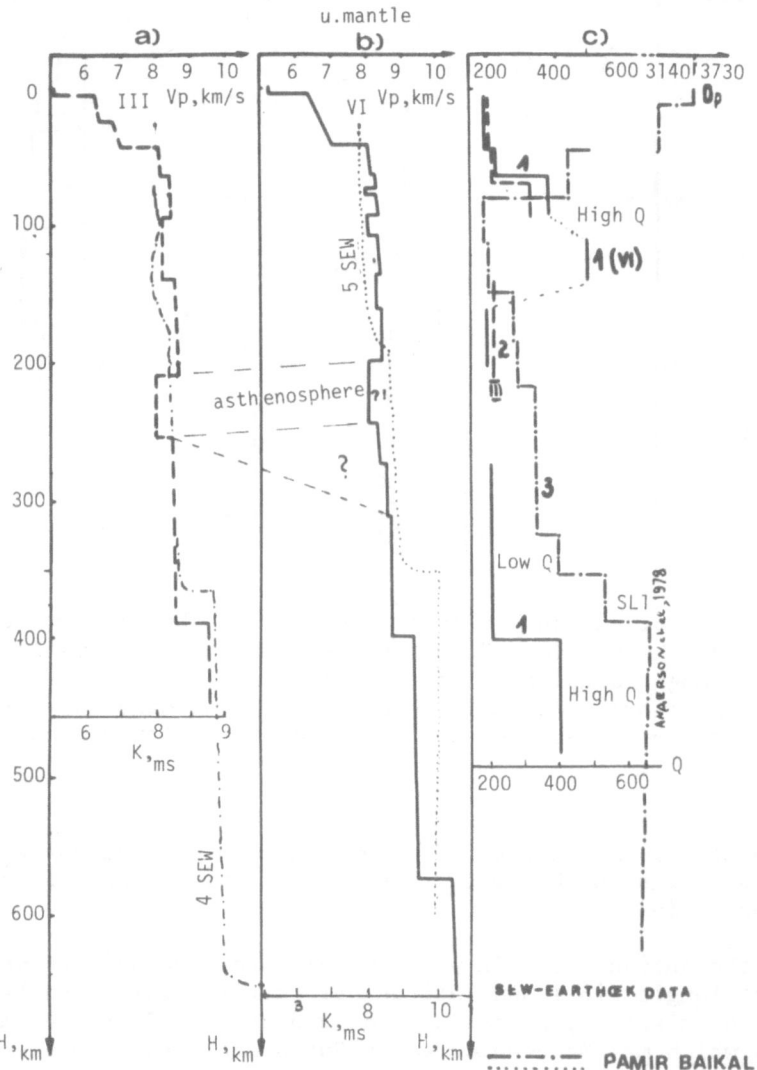

Fig. 11. One dimensional models of the upper mantle (profile III and IV at Siberian Platform. Figure 2) and Q factor curves. At a) and b) dashed lines correspond to EWS data obtained at Pamir-Baikal seismological profiles (see Figure 12 and 13). Note LVL at the depth 200 km which may be considered as the Gutemberg's asthenosphere.

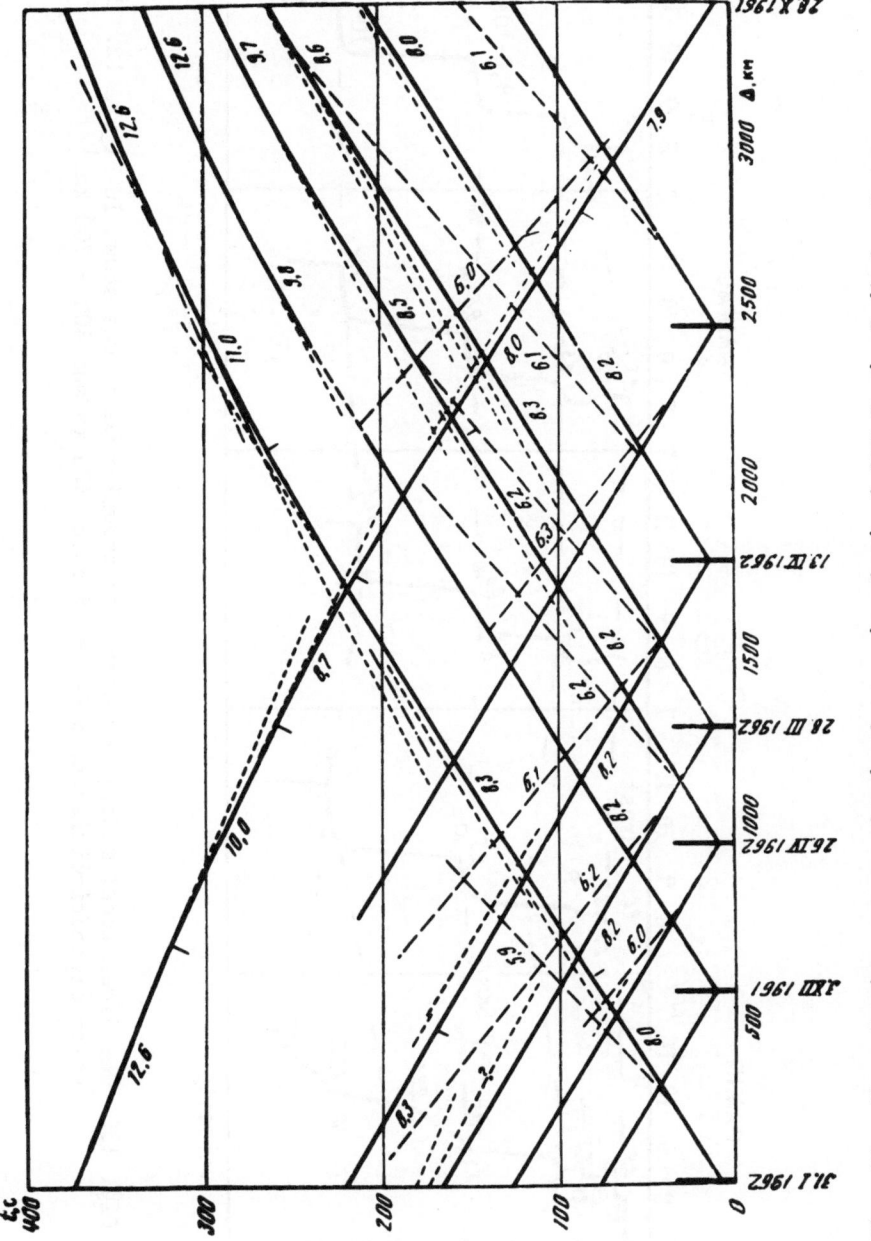

Fig. 12. Travel-time curves obtained at seismological LRP Pamir-Baikal. Dates correspond to the earthquakes epicenters. More than 100 record stations were distributed along the profile during about three years.[1]

109

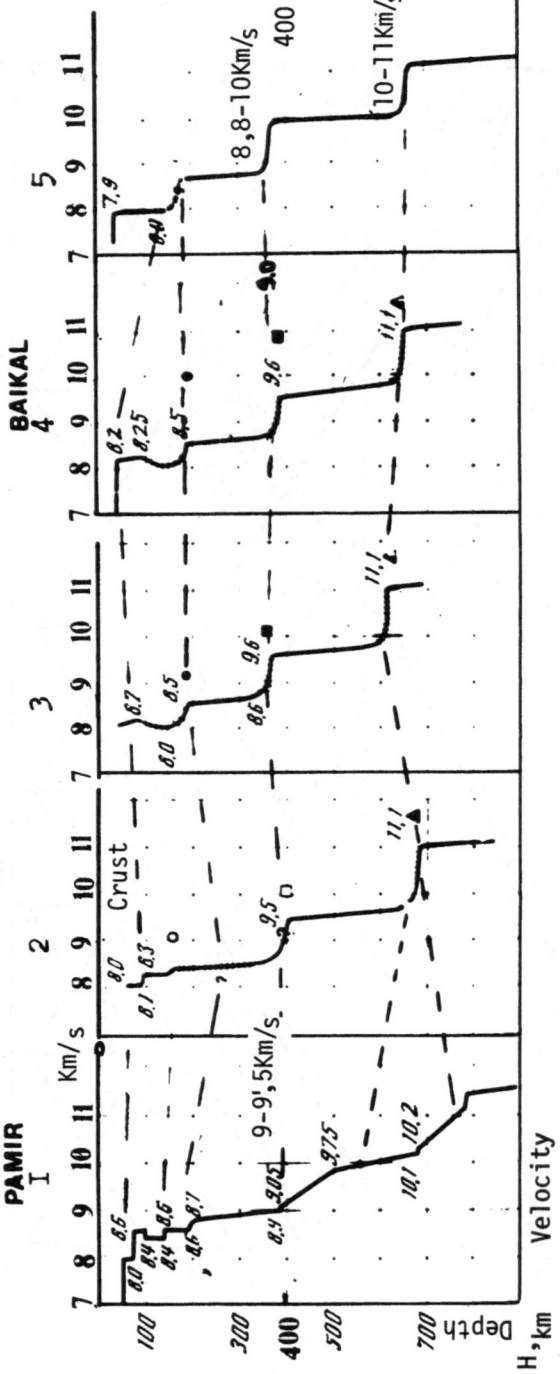

Fig. 13. One dimensional models correspond to travel-time curves shown in Figure 12. Note sharp discontinuities at a depth of about 400 km and 600 - 700 km.[1]

110

ACKNOWLEDGEMENTS

The author thanks Prof. R. Cassinis for the kind invitation to take part in the course of the "Ettore Majorana" International School of Applied Geophysics in March 1982, and Drs. A. V. Egorkin and N. I. Pavlenkova for assistance in review preparation.

REFERENCES

1. L. V. Antonova and F. F. Aptikayev, Experimental seismological study of the Earth's interior, A. M. Epinatieva, ed., <u>Nauka</u> 155 (in Russian), (1978).
2. I. P. Kosminskaya and N. K. Kapustian, Generalized seismic model of oceanic type. Physics of the Earth, <u>Izv. of Acad.Sci.USSR</u>, N 2, p. 35-45 (English translation), (1975).
3. Seismic models of the lithosphere for the major geostructures of the territory of the USSR, S. M. Zverev and I. P. Kosminskaya, eds., <u>Nauka. M.</u>, 184, (in Russian; abstract and contents in English), (1980).
4. S. M. Zverev and N. K. Kapustian, Seismic study of the Pacific ocean the lithosphere, <u>Nauka</u> 207 (in Russian), (1980).
5. N. I. Pavlenkova, Wave fields and crustal models (continental area), <u>Naukova Dumka, Kiev.</u>, p. 189 (1973).
6. L. P. Vinnik and A. V. Yegorkin, Wave fields and models of the lithosphere-asthenosphere from the data of seismic observation in Siberia, <u>Doklady of Acad.Sci.USSR</u>, 250, N 2:318-322 (1980).
7. A. V. Yegorkin and V. V. Kun, The P-wave absorption in the upper mantle. Physics of the Earth, <u>Izv.Acad.Sci.USSR</u>, N 4:25-36 (English translation), (1978).
8. A. V. Yegorkin and G. V. Egorkina, S waves in DSS observation. Geology and Geophysics N 6, <u>Novosibirsk.Nauka.</u>, 109-120 (1980).
9. I. P. Kosminskaya, Deep seismic sounding of the Earth's crust and upper mantle. Consultant bureau, New-York, London, 184 (1971).
10. I. P. Kosminskaya and N. I. Pavlenkova, Seismic models of the inner parts of the Euro-Asian continent and its margins, Sh. Keen, ed., <u>Tectonophysics</u>, 59:307-320 (1979).

MAIN FEATURES OF CRUSTAL STRUCTURES

IN MEDITERRANEAN COLLISIONAL ZONES

P. Giese

Institut für Geophysikalische Wissenschaften
Freie Universität Berlin
Rheinbabenallee 49
D-1000 Berlin 33

INTRODUCTION

In the Mediterranean region numerous examples of young collisional belts exist. Since more than hundred years these zones are intensively investigated by geological and petrological methods, and many data are available concerning the tectonics and the development of these mountain belts. The greatest recent progress in tectonic models in the Mediterranean region is the general acceptance, that since the early Mesozoic time large-scale horizontal movements must have taken place between Europe and Africa including smaller intermediate blocks. Such wide horizontal movements do not only govern the formation of the uppermost crust, but involve deeper horizons of the deeper crust and uppermost mantle as well. If discussing the structure of young collisional zones, geophysical investigations can contribute to solve problems concerning the structure and composition of deeper crustal levels.

Within the last two decades intensive geophysical studies have been carried out in the Mediterranean region. Among the various geophysical methods seismic studies have been widely applied. Sedimentary basins have mainly been explored by the seismic reflection technique, whereas the deeper crust and upper mantle have been investigated mainly using the seismic refraction method.

This paper aims to give a review of the main features of crustal structure of the Mediterranean collisional zones so far as deep seismic sounding data are available. Some tectonic considerations complete this study. This contribution is based on a large number of publications, the most recent ones are presented in the Geodynamics Series Vol. 7 "Alpine-Mediterranean Geodynamics", edited by BERCKHEMER and HSŪ (1982).

AREA UNDER STUDY AND SOME GENERAL ASPECTS

Fig. 1a shows the mountain ranges of the Mediterranean region and the position of the crustal profiles described in this paper. The activity was mainly directed to the Carpathians, Alps, Apennines, Dinarides, Hellenides, Pyrenees and to the Betic Cordillera. Fig. 1b presents the network of the seismic refraction lines existing up to 1980. In the sea areas exists a dense network of seismic reflection profiles with different penetration depths (MORELLI, 1975; MORELLI et al., 1975a,b,c; FINETTI et al., 1973; GIESE et al., 1982).

In seismic reflection work modern field and processing techniques produce excellent structural pictures. In refraction seismics the density of data is distinctly smaller, and consequently only the main structural features can be treated and displayed. Such a main feature is the crust/mantle boundary, the Mohorovičič-discontinuity. In the records of seismic-refraction measurements the waves overcritically reflected at the crust/mantle boundary are characterized by large amplitudes. Thus they can easily be detected in most of the seismogram sections.

Generally the crust/mantle boundary is regarded as an interface existing without any interruptions everywhere under the continents. It is defined by the level at which the velocity increases rapidly or discontinuously from crustal values of 6.0-6.8 km/s to upper-mantle values between 7.6 and 8.6 km/s. This strong velocity gradient is caused by the change from sialic to ultrabasic material. The above definition is applicable without problems in consolidated continental areas. But in the Mediterranean region with its young tectonics and wide horizontal displacements an anomalous behaviour of the crust/mantle boundary must be expected.

Two aspects are of importance. Some regions in the Mediterranean are characterized by anomalous high temperatures, and thus lower velocity values in the

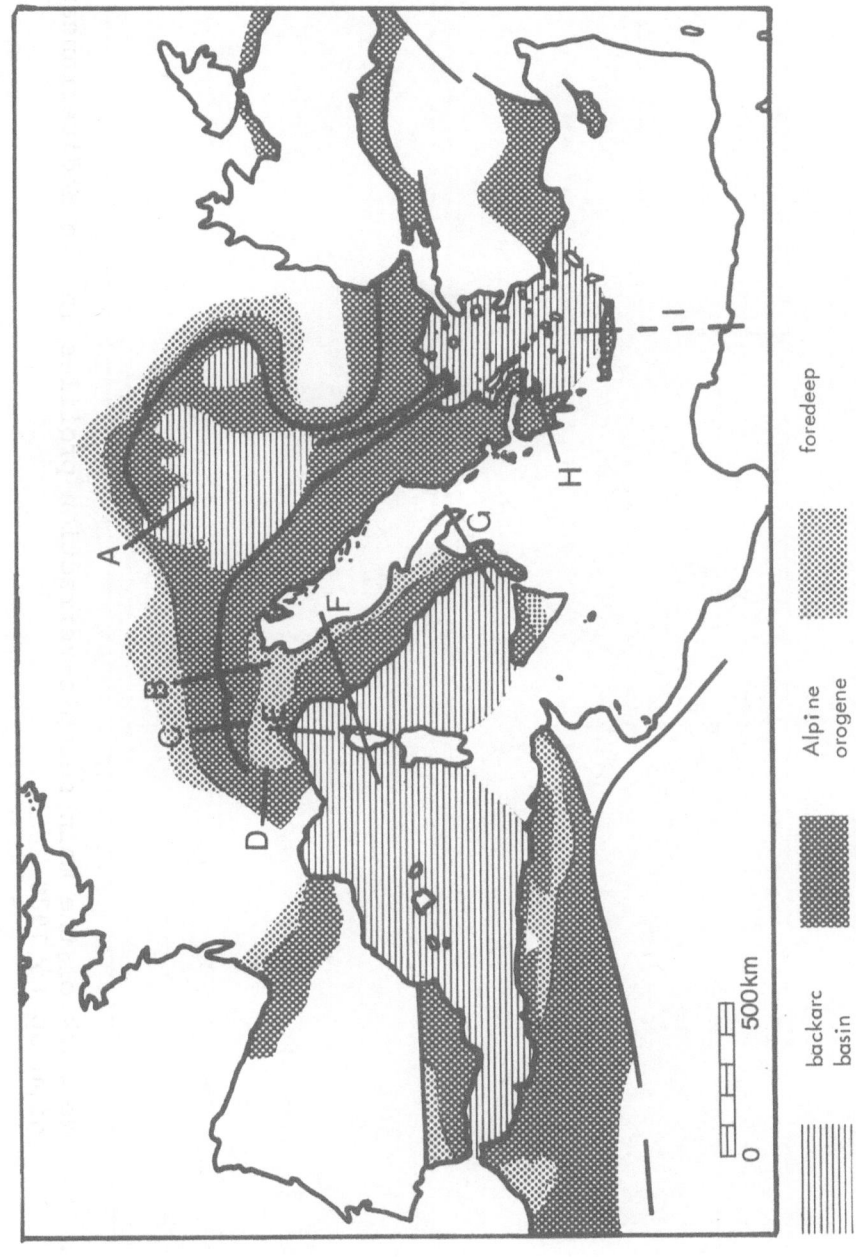

Fig. 1a. Area under investigation showing the main Alpine orogenes and the position of the crustal sections described in this paper

backarc basin

Alpine orogene

foredeep

0   500km

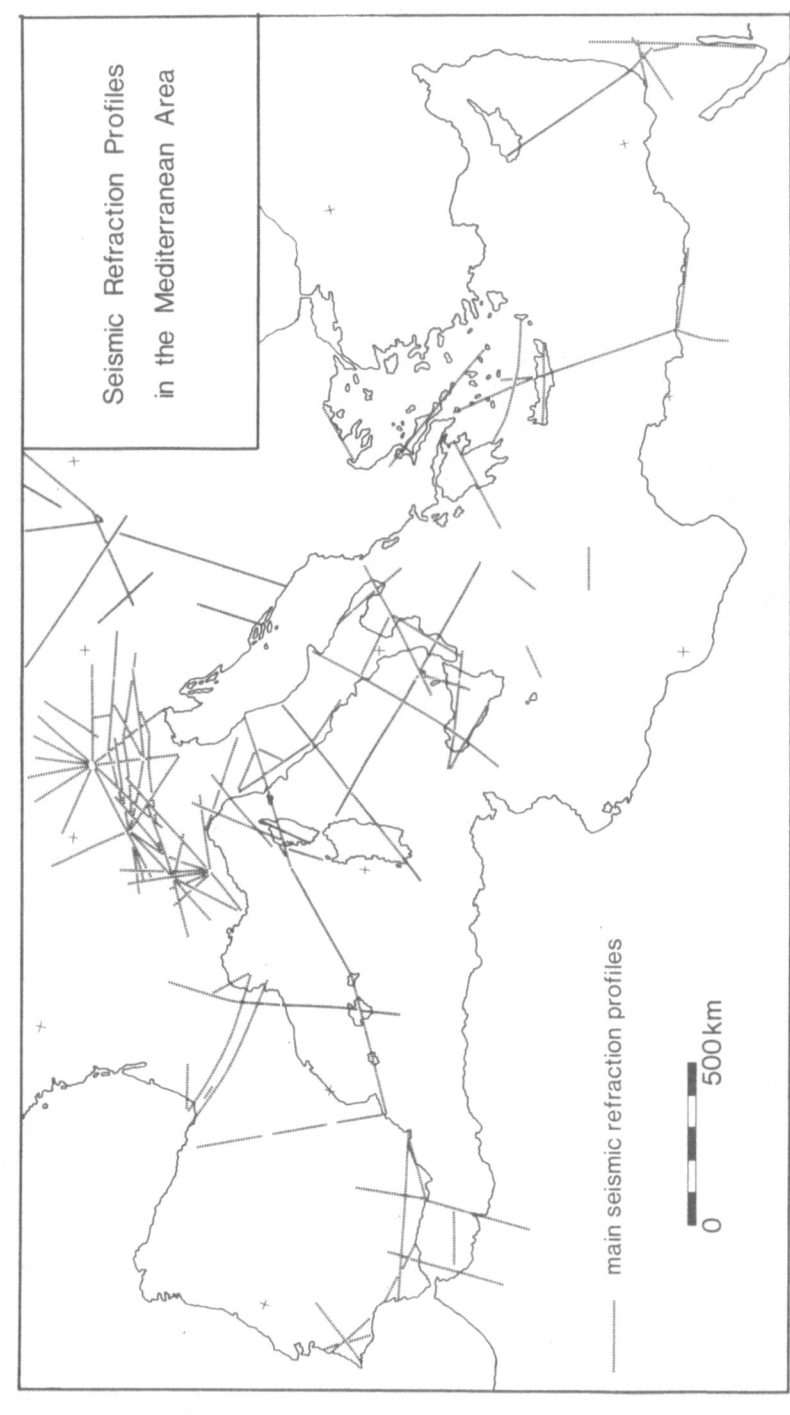

Seismic Refraction Profiles
in the Mediterranean Area

main seismic refraction profiles

0   500km

Fig. 1b. Network of the main seismic-refraction profiles in the Mediterranean region up to 1980

116

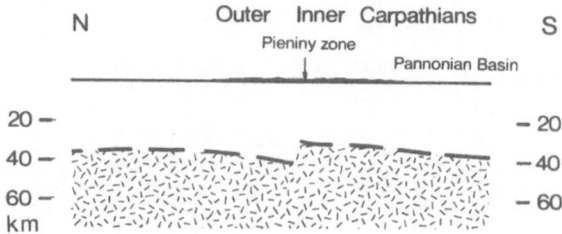

Fig. 2. Crustal cross section A
through Western Carpa-
thians. Note the change
in crustal thickness be-
neath the Pieniny zone.
Data from BERANEK et al.
(1972)

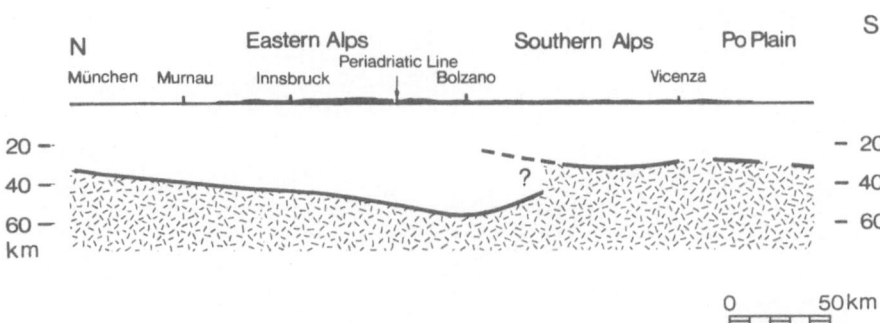

Fig. 3. Crustal cross section B through the Eastern
and Southern Alps. Surprisingly the maximum
crustal thickness is measured under the
Southern Alps. This feature can be explained
by crustal overlapping. Data from GIESE and
PRODEHL (1976), GIESE (1980), and MILLER et
al. (1982)

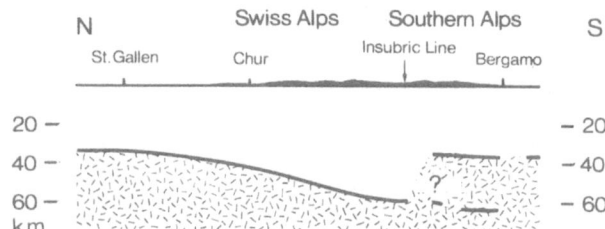

Fig. 4. Crustal cross section C through the Swiss Alps. Note the clearly expressed asymmetric structure. The position of the Insubric line coincides with the abrupt changes of crustal thickness. Data from GIESE and PRODEHL (1982), MÜLLER et al. (1980), and MILLER et al. (1982)

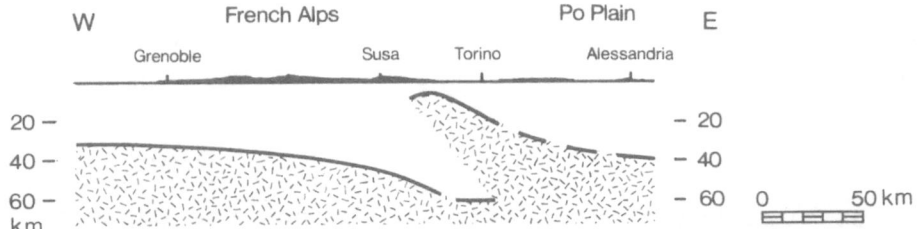

Fig. 5. Crustal cross section D through the Western Alps. Here the asymmetric structure is most pronounced. The crust of the Western Alps extends beneath upper-mantle rocks belonging to the Po-plain block. Data from GIESE and PRODEHL (1976), and MILLER et al. (1982)

crust and upper mantle must be taken into consideration.
So the velocity values within the strong-gradient zone
may be somewhat smaller than normal. In many zones the
strongest velocity gradient is found between 7.0 and
7.5 km/s whereas in the range between 7.5 and 8.5 km/s
only a weak increase is present. From petrological
point of view it is reasonable to associate the crust/
mantle boundary with that depth characterized by the
strongest velocity gradient within 7.0 and 8.5 km/s.

A second consideration concerning the structural
behaviour of the crust/mantle boundary is of importance
when studying tectonic relations between surficial
and deeper structures. Large-scale horizontal movements
require deep-seated shearing planes. Such shearing
planes can be assumed to exist within and at the base
of the sialic crust and in the uppermost mantle. The
emplacement of crustal units including basement rocks
and their thrusting over wide horizontal distances can
produce a stacking of high- and low-velocity layers.
The deeper crust and the uppermost mantle are involved
in such processes, too. Consequently portions of the
crust/mantle boundary may overlap each other.
A break-off of the crust/mantle boundary at shallow
depths and a continuation at a new, separated boundary
at greater depths is of importance for this study, and
attention must be focussed to such zones.
       The development of an orogene is characterized
by the migration of the tectonic and magmatic activity
from one side to the other side of an orogenic belt.
This direction of development defines the position of
the hinterland, where the activity starts and the position of the foreland where the activity dies out. It
will be outlined in the following, that the crustal
structure of the Mediterranean mountain belts is
strongly related to the couplet foreland-hinterland.

WESTERN CARPATHIANS

During extensive seismic crustal studies in SE-
Europe, profiles were observed through the Western and
Eastern Carpathians (BERANEK et al. 1972). From the
European foreland the crust/mantle boundary dips down
towards the axis of the Carpathians. In the Western
Carpathians beneath the Pieniny Klippen belt, separa-
ting the Outer from the Inner Carpathians, a crustal
thickness of 40 km has been measured. In the Eastern
Carpathians at the same tectonic position the thick-
ness of the crust is even near 50 km. Along the
Pieniny Klippen belt the crustal thickness reduces

abruptly from 40-50 km to 25 km (Fig. 2). The thickness remains rather constant across the Inner Carpathians and Pannonian Basin. Thus it can be stated that the hinterland shows a thin crust, whereas the outer belt lying on the European foreland is characterized by a thick crust. The Pieniny Klippen zone separates these two crustal types.

EASTERN ALPS

Reviews on the structure and the geological evolution of the Eastern Alps were published recently e.g. by ANGENHEISTER et al. (1975), BÖGEL et al. (1976) and OBERHAUSER (1980). Seismic crustal studies were started in the Alps during the early fifties. Reviews on the results were published by GIESE and PRODEHL (1976), GIESE (1980), GIESE et al. (1982), MILLER et al. (1982).

The section Murnau-Innsbruck-Bolzano-Verona can be seen as typical for the crustal structure of the Eastern Alps. Generally the crust/mantle boundary dips down from the foreland towards the axis of the Alps (Fig. 3). The maximum depth (55-60 km) is reached south of the Periadriatic Line near Bolzano. At the southern border of the Alps the crust shows a thickness near 30 km. If tracing this boundary northwards it dies out S of Bolzano at 20-25 km depth. This interpretation implies a wedge-like shaped Adriatic crust thinning from the southern margin of the Alps towards the Periadriatic Line. A deeper crustal portion follows under this Adriatic crust with a second crust/mantle boundary at its base belonging to the European block. From tectonic point of view this lower crustal portion must be composed of crustal fragments of Austroalpine units and/or intra-geosynclinal ridges. More details on internal structure of crust and relation to other geophysical parameters are given by MILLER et al. (1982).

WESTERN ALPS

A crustal cross section passing the Swiss and Bergamasc Alps between Chur and Bergamo demonstrates very distinctly the asymmetric structure of the Alpine crust (Fig. 4). At the Insubric Line the crustal thickness changes abruptly from 55 km N of this suture to 35 km S of it. There are some weak indications in the seismic record sections for a very deeply seated discontinuity (ca. 70-75 km) under the Southern Alps which may be linked with the deep crust/mantle boundary in the central zone of the Alps. The problem of high- and low-velocity layers and their tectonic implications are discussed by MÜLLER et al. (1980).

Passing the Western Alps a very strange picture is met between Grenoble and Torino (Fig. 5). This profile crosses the Ivrea gravity-high at the inner side of the Western Alps and the southern end of the geological Ivrea zone characterized by its outcropping basic and ultrabasic rocks. The western part of this section looks quite normal, the crustal thickness increases from ca. 30 km in the foreland to about 60 km under the central zone of the Western Alps. Immediately E of the Periadriatic Line,which at its western end is named Insubric Line, high-velocity rocks (7.2 km/s) were detected at or near the surface along the axis of the Ivrea gravity-high. In the northern part of the gravity-high between Ivrea and Locarno these high-velocity/high-density rocks can be associated with outcropping basic and ultrabasic rocks. These anomalous material must extend down to about 20 km (NIGGLI, 1946; GIESE, 1968). Under this high-density/high-velocity material exists a very pronounced low-velocity zone (4-5 km/s). The crust/mantle boundary measured in a depth of 50-60 km under the central zone of the Western Alps continues eastwards under the gravity-high and extends even under western margin of the Po-Plain. The so-named geophysical Ivrea body continues at its eastern flank into the Adriatic microplate. Here an overlapping of two tectonic different crustal units is clearly demonstrated. The geological problems of this strange Ivrea zone are discussed by AHRENDT (1972, 1980) and at 2nd Symposium "Ivrea-Verbano" held at Varallo/Italy, in 1978.

Summarizing the results obtained in the Carpathians and the Alps the following can be concluded: In both orogenes the crustal structure is asymmetric. From the foreland towards the central zone of the mountain belt the crust/mantle boundary dips down more or less continuously. A distinct change in crustal structure takes place in the most internal zone where a prominent suture (Pieniny Klippen Zone and Peridariatic Line) separates the central zone from the hinterland. Passing this tectonic lineament a sudden reduction of crustal thickness for the hinterland crust is observed. Comparing the three typical cross sections through the Alps (Fig. 3, 4 and 5) it can be seen that the crust of the northern border of the hinterland, the Southern Alps, changes systematically its structure. In the eastern section upper-mantle material thins out at the base of the crust, in the middle section it reaches the Periadriatic line, whereas in the western section lower crustal/upper-mantle rocks reach the surface.

APENNINES

Due to the activity of Italian colleagues the crust of the Italian penninsula has been studied very intensively by seismic refraction measurements (GIESE et al., 1976, 1982; MORELLI et al., 1977; CASSINIS, 1979, 1981; NICOLICH, 1981).

The Northern Apennines on one side and the southern part of the Western Alps and their prolongation to Corsica on the other side are part of the same orogenic belt but with opposite tectonic developments. The polarity of the Western Alps is directed towards Europe whereas that of the Apennines show towards the Adriatic microplate. Thus, if going from the Western Alps into the Northern Apennines the function of the Adriatic microplate changes from a hinterland in a foreland.

Two sections describe the crustal structure in the Northern Apennines. The section passing the Ligurian Apennines (Fig. 6) shows a crust of the Adriatic microplate getting continuously thicker from the Ligurian Sea towards the Po-Plain. The next section, Fig. 7, runs from the Balearic Sea to the Adriatic Sea, passing Corsica, Elba, Tuscany and Umbria.

The crust of the Balearic Sea looks typically oceanic, the crust/mantle boundary is found at a depth of 11 km (HIRN et al., 1976). The second portion of this profile, the Corsica-Sardinia block, presents a typical continental crust with a thickness of 30 km. The third region comprises the Ligurian shelf and Tuscany which belong from tectonic point of view to the Adriatic microplate. Under Elba a well expressed crust/mantle boundary is found at only 20-25 km depth which seems to be separated from that of the Corsica block. This shallow boundary continues at about the same depth level up to the border region between Tuscany and Umbria. This thin crust agrees well with the concept that the Tuscany region presents a back-arc basin being in an early stage of development. The fourth part of this profile crosses the Umbria-Marches arc which shows a Bouguer gravity minimum and a crustal thickness of 35 km. Near Ancona at the Adriatic coast the crustal thickness gets thinner to 30 km.

A special and complex structure must be expected between the Corsica block being a European fragment and Elba as the western margin of the Adriatic microplate. As indicated in the seismic record sections the

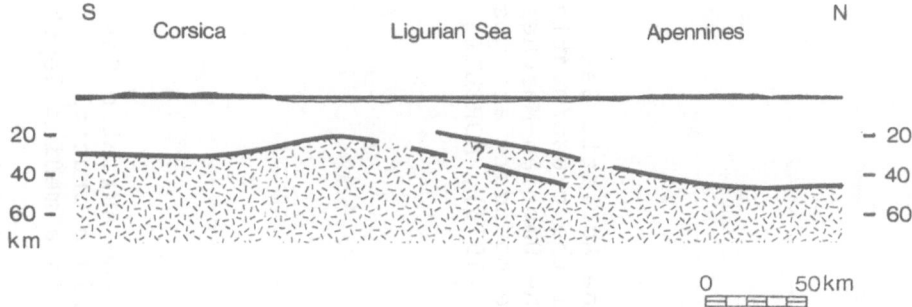

Fig. 6. Crustal cross section E between Corsica and
         the Ligurian Sea. Here exists a complicated
         structure with intercalating crustal and
         upper mantle fragments. Data from MORELLI
         et al. (1977)

crust/mantle boundary of the Corsica block seems to
plunge eastwards under the shallow crust/mantle boundary of the Adriatic microplate. A similar structure was
detected on profiles between Corsica and the Ligurian
Apennines.

Southern Italy and the adjacent sea areas have
been investigated by extensive seismic measurements.
A cross section describing the main features of crustal
structure in that region is presented in Fig. 8. The
deep-sea part of the Tyrrhenian Sea shows a typical
oceanic structure with a crustal thickness of 11 km. In
contradiction the Tyrrhenian shelf including a narrow
coastal zone has a thinned continental crust with the
crust/mantle boundary at 18-20 km depth. Going onshore
the crustal thickness increases jump-like to 40-45 km.
Towards E and NE the crust gets thinner and under
Gargano and Apulia as the foreland of the Southern
Apennines a crustal thickness of 30 km exists. Here
Mesozoic and Tertiary sediments with a thickness of
10-15 km are proved. Consequently a thinned basement
crust must exist here (15-20 km).

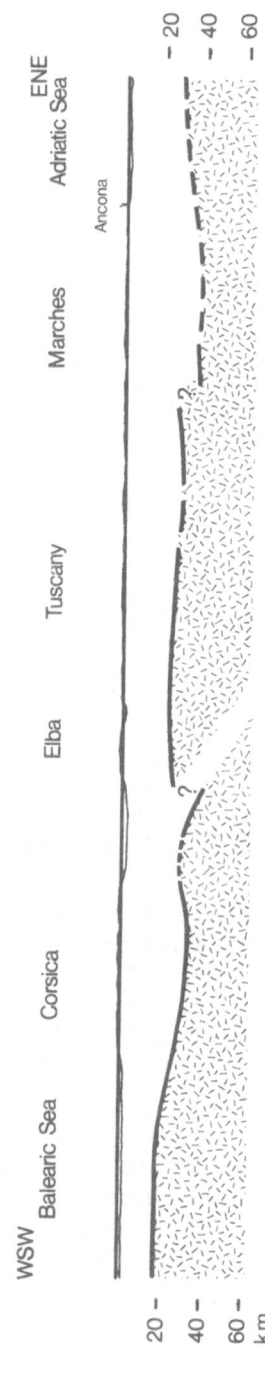

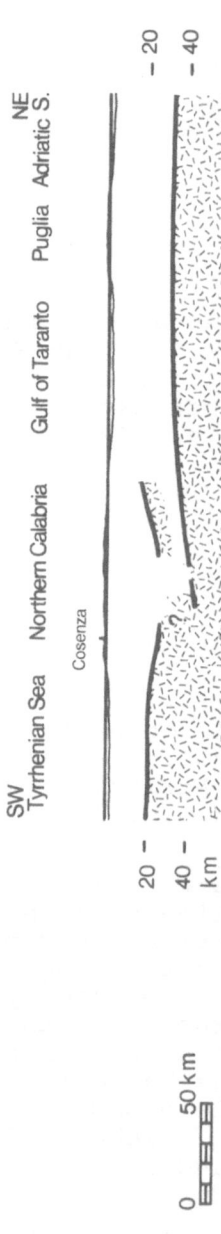

Fig. 7. Crustal cross section F between the Balearic Sea and the Adriatic Sea passing Corsica and Tuscany. A complicated structure of the crust/mantle boundary is met between Corsica and Elba and beneath the Umbrian-Marches arc. Displacements and overlappings of the crust/mantle boundary are possible here. Data from GIESE et al. (1982), HIRN et al. (1976), MORELLI et al. (1977)

Fig. 8. Crustal cross section G between the Tyrrhenian Sea and the Adriatic Sea passing Northern Calabria and Puglia. Beneath NE-Calabria a crustal stacking is evidenced by a high-velocity body situated in a medium depth level. Data from GIESE et al. (1982)

On the seismic refraction profile Paestum-Gargano a discontinuity at a depth of 60 km could be detected 45 km offshore in the Tyrrhenian Sea. This boundary is associated with that found at 45 km depth under the central axis of the Apennines. This correlation again implies a crustal doubling with a westwards dipping shear-plane. In principle the same crustal structure as in the Southern Apennines exists below the northern coast of Sicily.

A comparison of the Apenninic cross sections shows that there is a systematic change of structure from N to S. The thin Tuscany crust continues into that of the Tyrrhenian shelf, whereas the thick crust in the Umbria-Marches arc can be followed up into the central zone of the Southern Apennines.

HELLENIDES AND DINARIDES

The Hellenides and Dinarides form the structural backbone of the Balkan peninsula between Northern Jugoslavia to Anatolia. Information on crustal structure of the Hellenides are available from the Peloponnesus, the Aegean Sea and Crete (MAKRIS, 1977, 1978). The western coast of the Peloponnesus shows surprisingly a thin crust tectonically acting as foreland (Fig. 9). This thin crust continues for 20-30 km towards the inland. Then the shallow crust/mantle boundary of the foreland disappears suddenly, and a second, separated boundary being of 45-50 depth forms the base of the crust. This thick crust is proved by a Bouguer minimum, too. If going eastwards into the internal zones of the Hellenides the crustal thickness decreases to 25 to 30 km.

Although the orogenetic polarity of the Hellenides points from E to W, this part of the Hellenides does not correspond to the typical model of the Alps and Carpathians. In order to explain the thick crust in the external zone and the abrupt change of thickness under the foreland, an underthrusting of eastern crustal fragments beneath the crust of the foreland is suggested.

A NS-section through the southern Aegean islands and Crete resembles to that which passes through the Western Carpathians (Fig. 10). The sea of Crete, acting now as hinterland, shows a thinned crust with

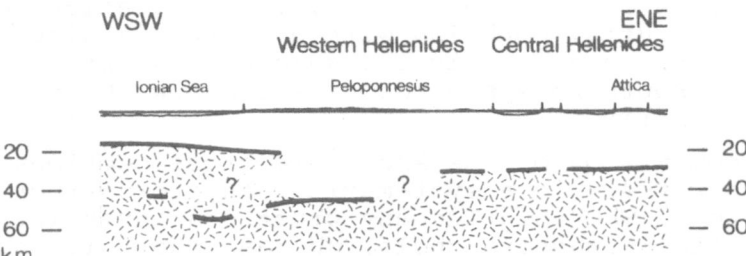

Fig. 9. Crustal cross section H through the Pelepon-
         nesus. Here the most striking feature concerns
         the course of the crust/mantle boundary be-
         neath the foreland. In contradiction to the
         other examples here the foreland crust shows
         a sudden change in crustal thickness. Data
         from MAKRIS (1977, 1978), and GIESE et al.
         (1982)

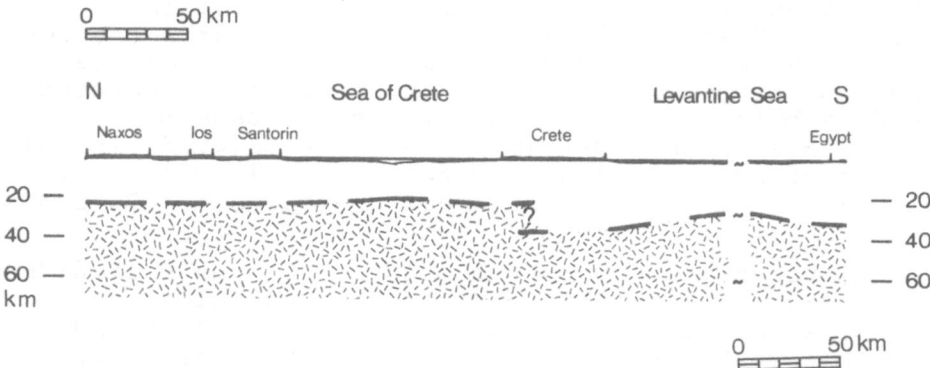

Fig. 10. Crustal cross section I between the Aegean Sea
         and North Africa. A sudden change in crustal
         thickness takes place at the northern coast of
         Crete. Here again the abrupt change coincides
         with the zone where the internal zone of the
         Hellenides borders to the external chain. Data
         from MAKRIS (1977, 1978) and GIESE et al. (1982)

20-22 km. Under the northern coast of Crete an abrupt increase of crustal thickness to 30-40 km takes place. This zone coincides with the boundary between the northern region deformed in Late Eocene and a Southern area showing a Miocene deformation. Between Crete and the African coast crustal thickness is near to 26 km whereas along the Egyptian coast the crust/mantle, boundary is at 30 km depth (WEIGEL, 1979)

Two seismic refraction profiles are available which traverse the Dinarides (BERANEK et al., 1972). Here again the Inner Dinarides as internal zone show a crust being only 25 km thick, whereas in the External Dinarides a crustal thickness of 50 km was measured. In the Adriatic Sea, the foreland of the Dinarides, a crustal thickness of 30 km can be expected.

THE BETIC CORDILLERA

The Betic Cordillera shows a vergency which is directed from S to N towards the Spanish Meseta. The most internal zones are situated along the southern coast of Spain. Here along this coast crustal thickness measures 23-24 km. Going northwards into the central zone of the Betics the thickness increases to about 40 km. (BANDA et al., 1980). The foreland of the Spanish Meseta shows a normal thickness of 30-32 km. Summarizing these results it can be stated that the Betics present again the typical asymmetric structure with a thin crust in the internal zone.

The Pyrenees contrast distinctly in their structure and development with other Mediterranean orogenes. The simplex schematization consists of an axial zone bordered by two marginal troughs. A wide oceanic trough never exists here. The seismic data obtained in the Pyrenees are still under interpretation. The first results show the following picture (GALLART et al., 1980): On the French side of the Pyrenees, in the north Pyrenean zone, the crustal thickness is between 28 and 34 km. A distinct crustal thickening of at least 10 km exists in the Paleozoic Axial zone on the Spanish side. Near the Mediterranean coast the crust turns out to ca. 20-22 km. Although the structure is asymmetric, too, a comparison with the other Mediterranean mountain belt is difficult, due to the different geological developments of the Pyrenees.

CONCLUSIONS

Generation of oceanic basins, consumption and subduction of these basins and collision of large as well
as small continental blocks and fragments are characteristic for the geotectonic processes in the Mediterranean. All mountain belts represent zones of strong
shortening and compression combined with a clear polarity of several geological phenomena. After the consump-
ion of oceanic areas two continental elements get contact and can collide. The comparison and analysis of
crustal cross sections traversing a number of Mediterranean mountain chains demonstrate two types of crustal collisional structures (GIESE and REUTTER, 1978),
(Fig. 11).
(1) The colliding crusts overlie each other synthetic-
    ally in respect to the thrust planes of the nappes
    and the preceding subduction direction.
(2) In the other case they overlie each other anti-
    thetically to these movement planes.
In case of a synthetic collision a continental block of
an upper plate (named hinterland) with its more or less
complete crust and adherent parts of the uppermost
mantle overrides a downgoing foreland crust and intermediate continental units, previously situated as ridges between the two colliding continental blocks. This
type of collision is met in the Carpathian, the Alps,
the Southern Apennines and probably in the Betic Cordillera.

In the other case, the antithetic one, the frontal
edge of the hinterland crust is pushed beneath or within the crust of the foreland. Here a crustal delamina-
tion, detachment or splitting must be required (OXBURGH,
1972; BIRD, 1978; REUTTER et al., 1980). This delamina-
tion allows an uplift of the internal and upper parts
of the foreland crust, whereas its deeper portions
must sink down. Examples for this case can be found in
the Northern Apennines and the Hellenides.

Fig. 11.

Two sections show-
ing the models of
synthetic and an-
tithetic crustal
collisional struc-
tures

REFERENCES

Ahrendt, H., 1972, Zur Stratigraphie, Petrographie
    und zum tektonischen Aufbau der Canavese-Zone und
    ihrer Lage zur Insubrischen Linie zwischen Biella
    und Cuorgné (Norditalien), Göttinger Abh. Geol.
    Paläont. 11, 89 S.
Ahrendt, H., 1980, Die Bedeutung der Insubrischen
    Linie für den tektonischen Bau der Alpen, N. Jb.
    Geol. Paleont., 160, 3, 336-362.
Angenheister, G., Bögel, H., Gebrande, H., Giese, P.,
    Schmidt-Thome, P., Zeil, W., 1972, Recent investi-
    gations of surficial and deeper crustal structure
    of the eastern and southern Alps, Geol. Rdsch.
    61,349-395.
Angenheister, G., Bögel, H., Morteani, G., 1975, Die
    Ostalpen im Bereich einer Geotraverse vom Chiem-
    see bis Vicenza, N. Jb. Geol. Paläont., Abh. 148,
    50-147.
Banda, E. and Ansorge, J., 1980, Crustal structure
    under the central and eastern part of the Betic
    Cordillera, Geophys. J.R. astr. Soc., 63, 515-532.
Beránek, B., Weiss, J., Hrdlicka, A., Dudek, A., Zoun-
    kova, M., Suk, M., Feifar, M., Militzer, H.,
    Knothe, H., Mituch, E., Posgay, K., Uchmann, J.,
    Sollogub, V.B., Chekunov, A.V., Prosen, D., Milo-
    vanovič, B., Roksandic, M., 1972, The results of
    the measurements along the international profiles,
    in: The Crustal Structure of Central and South-
    eastern Europe Based on the Results of Explosion
    Seismology Geophs., Transaction Budapest, 133-148.
Berckhemer, H. and Hsü, K (eds), 1982, Alpine-Mediter-
    ranean Geodynamics, Geodynamics Series Vol. 7,
    AGU, Washington, 216 p.
Bird, P., 1978, Initiation of intracontinental sub-
    duction in the Himalaya, J. Geophys. Res., 83,
    4975-4978.
Bögel, H., Schmidt, K., 1976, Kleine Geologie der Ost-
    alpen, Ott, Thun, 231 p.
Cassinis, R., Franciosi, R., Scarascia, S., 1979, The
    structure of the Earth's crust in Italy - A pre-
    liminary typology based on seismic data, Boll. di
    Geof. Teor. ed appl., XXI, 82, 105-126.
Finetti, J. and Morelli, C., 1973, Geophysical explo-
    ration of the Mediterranean Sea. Boll. Geofis.
    teor. ed appl. 15, 263-341.
Gallart, J., Daignieres, M., Banda, E., Surinach, E.,
    Hirn, A., 1980, The Eastern Pyrenean domain, la-
    teral variations at crust-mantle level, Ann.

Geophys., 36, 2, 141-158.

Giese, P., 1968, Die Struktur der Erdkruste im Bereich der Ivrea-Zone, Schweiz. Min. Petr. Mitt. 48, 261-284,

Giese, P., 1980, Krustenstruktur der Alpen - Ein Überblick verbunden mit einigen tektonischen Betrachtungen, Berliner Geow. Abh. (A), 20, 51-64.

Giese, P. and Prodehl, C., 1976, Main features of crustal structure of the Alps, in: P. Giese, C. Prodehl, A. Stein (Eds.), Explosion Seismology in Central Europe, Springer-Verlag Berlin-Heidelberg-New York, 347-376.

Giese, P. and Reutter, K.-J., 1978, Crustal and structural features of the margins of Adria microplate, in: H. Closs, D. Roeder and K. Schmidt (Eds.), Alps, Apennines, Hellenides, Schweizerbart, Stuttgart, 565-588.

Giese, P., Reutter, K.-J., Jacobshagen, V., Nicolich, R., 1982, Explosion seismic crustal studies in the Alpine-Mediterranean Region and their implications to tectonics processes, in: Alpine-Mediterranean Geodynamics, edited by H. Berckhemer and K. Hsü, AGU Washington, Geodynamics, Series Vol. 7, 39-74.

Hirn, A. and Sapin, M., 1976, La croûte terrestre sous la Corse: données sismiques, Bull, Soc. geol., France, no. 5, 1195-1199.

Makris, J., 1977, Geophysical investigations of the Hellenides, Hamburger Geophys. Einzelschriften, H. 27, 98 p

Makris, J., 1978, A geophysical study of Greece based on: deep seismic soundings, gravity, and magnetics, in: Alps, Apennines, Hellenides, edited by H. Closs, D. Roeder, and K. Schmidt, Schweizerbart, Stuttgart, 392-400.

Miller, H., Müller, St., Perrier, G., 1982, Structure and Dynamics of the Alps - A geophysical inventary in: Alpine-Mediterranean Geodynamics edited by H. Berckhemer and K. Hsü, AGU Washington, Geodynamics, Series Vol. 7, 175-204.

Morelli, C., 1975, Geophysics, of the Mediterranean, Newsletter of the cooperative investigations in: the Mediterranean, no. 7, Monaco, 29-111

Morelli, C., Gantar, C., Pisani, M., 1975a, Bathymetry, gravity and magmatism in the strait of Sicily and in the Jonian Sea, Boll. Geof. Teor. ed appl. XVII, 65, 39-58

Morelli, C, Pisani, M., Gantar, C., 1975b, Geophysical studies in the Aegean Sea and in the Eastern

Mediterranean, Boll. Geof. Teor. ed appl. XVII,
65, 127-168

Morelli, C., Pisani, M., Gantar, C., 1975c, Geophysical anomalies and tectonics in the Western Mediterranean, Boll. Geof. Teor., ed appl. XVII, 65,
211-249

Morelli, C., Giese, P., Carrozzo, M.T., Colombi, B.,
Guerra, I., Hirn, A., Letz, H., Nicolich, R.,
Prodehl, C., Reichert, C., Röwer, P., Sapin, M.,
Scarascia, S., Wigger, P., 1977, Crustal and upper
mantle structure of the Northern Apennines, the
Ligurian Sea, and Corsica, derived from seismic
and gravimetric data, Boll. Geof. Teor., ed appl.
75-76, 199-261

Mueller, St., Ansorge, J., Egloff, R., Kissling, E.,
1980, A crustal cross section along the Swiss
Geotraverse from the Rhinegraben to the Po Plain,
Eclogae geol. Helv., 73/3, 463-483.

Nicolich, R., 1981, Crustal structures in the Italian
Peninsula and surrounding seas: a review of DSS
data, in: Sedimentary Basins of Mediterranean
Margins, edited by F.C. Wezel, C.N.R. Italian
Project of Oceanogr., 3-18 pp.

Niggli, E., 1946, Über den Zusammenhang zwischen der
positiven Schwereanomalie am Südfuß der Westalpen
und der Gesteinszone von Ivrea, Eclogae geol.
Helv. 39, 211-220.

Oberhauser, R., (Hrsg.), 1980, Der geologische Aufbau
Österreichs, Wien-New York, 695 S.

Oxburgh, E.R., 1972, Flake tectonics and continental
collision, Nature, 239, 202-204

Reutter, K.-J., Giese, P., Closs, H., 1980, Lithospher
ic split in the descending plate: observations
from the Northern Apennines, Tectonophysics 64,
T1-T9, Elsevier Scientific Publishing Comp. Am-
sterdam.

Weigel, W., 1979, Seismisches Projekt Ägypten/Kreta,
SPEC 1978 (abstract), 39. Jahrestagung der Deut-
schen Geophys. Ges. Kiel, April 1979.

POTENTIAL AND CONSTRAINTS OF NEAR VERTICAL REFLECTION SEISMICS IN

NON-SEDIMENTARY DEEP FORMATIONS

K. Helbig   and   J. Schmoll

Vening Meinesz Laboratory   formerly
Rijksuniversiteit Utrecht   Prakla-Seismos GmbH
P.O. Box 80.021   P.O.Box 510530
3508 TA Utrecht   3000 Hannover
The Netherlands   W.-Germany

INTRODUCTION

Without further specification, the word 'deep' in connection
with a seismic survey has no precise meaning: applied loosely, it
means just 'deeper than most surveys'. A seismic section or a
seismogram without time scale contains no clue to the depth of the
reflectors or the total time. As a matter of fact, the conceptual
difficulties to expand the state-of-the-art to shallower reflections
are just as large as those for the expansion to deeper reflections.
That does not mean that the technological difficulties are of comparable magnitude: the main message of this article is that nearly
everything in exploration seismics can be scaled. But expansion to
greater depth requires larger sources with greater total energy,
greater offsets with longer cables, and larger receiver groups.
Such changes are more expensive than a change to smaller scale.

Since this volume contains an excellent case history of current
reflection seismic probing of the lower crust, we concentrate on
basic principles of reflection seismics, on potential improvements
of data processing to assist in unravelling complex structures.
Changes in the acquisition technology are of much greater significance than data processing steps: while processing steps can be
tried at any time (provided the tapes are still available and the
project has access to a computer), there is no substitute for good
acquisition technology. Post-facto adjustments for too narrow band-
width, inaccuracies in source-and receiver geometry, and static
corrections are so complicated to be sometimes impossible.

While most parameters can be scaled, there is one principal difference between a reflection seismic survey aimed at the lower crust and the surveys for which most of the current technology has been developed and which set the technical standard against which, in the final analysis, success is measured: the sedimentary section of the upper crust is typically layered, in many places cyclically, while the lower crust is expected to be dominated by intrusive bodies like laccoliths and batholiths. At first thought it might seem likely that such bodies have strongly curved - and generally rather arbitrarily oriented - boundaries. However, abundant deep reflections from subhorizontal, subplanar boundaries indicate that this is not the case. It stands to reason that gravitational control constrains such boundaries to a nearly horizontal attitude.

BASIC PRINCIPLES OF REFLECTION SEISMICS

1. The seismic trace

A seismic trace is a (continuous or discrete) time function depending on source and receiver. Both location and nature of source and receiver have an influence on the trace: source and receiver generally consist of fixed patterns. The prime purpose of these patterns is a discrimination against (horizontal) wave numbers that are smaller than the dimensions of the patterns. In this way the influence of surface waves - particularly source-generated surface waves - is reduced. However, these patterns result also in an angular restriction, so that essentially only reflectors below a certain dip contribute to the seismogram.

Phase

A seismic trace can conveniently be regarded as the real part of a complex function with continuous phase and amplitude:

$$(1) \quad p(t) + i\,\tilde{p}(t) = p_{inst}(t) \cdot (\cos\varphi_{inst}(t) + i \cdot \sin\varphi_{inst}(t)), \text{ where}$$

$$p(t) = \text{Re}\left[p_{inst}(t) \cdot (\cos\varphi_{inst}(t) + i \cdot \sin\varphi_{inst}(t)) \right]$$

$$\tilde{p}(t) = \text{Jm}\left[p_{inst}(t) \cdot (\cos\varphi_{inst}(t) + i \cdot \sin\varphi_{inst}(t)) \right], \text{and conversely}$$

$$p_{inst}(t) = \sqrt{p^2(t) + \tilde{p}^2(t)} \text{ and } \varphi_{inst}(t) = \tan^{-1}(\tilde{p}(t)/p(t)) + n \cdot \pi$$

The imaginary part $\tilde{p}(t)$ of (1) - the so-called 'allied signal' - can easily be obtained: one either applies the Hilbert-transform to the trace, or one Fourier-transforms the trace, augments the phase by 90 degrees, and transforms back into the time domain (see

Taner, Koehler and Sheriff, 1979) back into the time domain. Here we are not concerned with the details of the process but only with the fact that there is - in principle - for each instance a unique phase value.

Amplitudes

The amplitude of the seismic trace concerns us for two reasons: firstly, the amplitude at 'target time' must be high enough so that the signal can be detected above the ubiquitous seismic noise; in order to achieve that, one must overcome the amplitude decrease due to spherical spreading, to absorption, and to transmission losses. Secondly, just because of the unavoidable amplitude decrease, the amplitude at the beginning of the record must be much higher than at target time. One thus needs instruments that can cope with an amplitude range of one to a million or better without saturation effects or less of significant information.

2.   Seismic events

A seismic event is a line-up of phases across a 'family' of seismic traces. Family is used loosely for a set of traces with parameters - source - and receiver-coordinates - that follow a simple rule, e.g. a common source and receiver at increasing distances. In order for the event to be visible, the amplitude must be high enough, but also the phase-noise throughout the family must be sufficiently small. Phase fluctuations can, of course, be considered as time fluctuations. However, a time shift of ten milliseconds would completely destroy a phase line-up at 100 Hz, while it could easily be tolerated at 10 Hz.

Phase consistency is obtained under two conditions:
(i) there are no drastic changes in the ray-geometric parameters from trace to trace within the family; (ii) time fluctuations due to near-surface irregularities have been removed as 'static corrections'. The second requirement could be considerably relaxed if traces would be very close together, since even near-surface effects rarely occur abruptly. However, such an acquisition geometry would be highly wasteful: the minimal trace spacing should be controlled by the lateral resolution, not by surface conditions.

Seismic events are the raw material of interpretation: no matter how sophisticated the later processing is, the final interpretation is based on the recognition of events in the seismic section. There are three main classes of seismic events, distinguished by the type of raypath that gave rise to their occurence:
(i) reflections - where the raypath makes an acute angle with the interfaces, (ii) refractions - where the ray is, for a section of the path, critically refracted and follows an interface, and

(iii) diffractions; these are distinguished from reflections by
the fact that the inhomogeneity that caused their return to the
surface is - at least in one (near-) horizontal direction - smaller
than the wave length of the signal.

Besides these three classes of events there are phase line-ups
that generally are classed as noise: surface waves generated by the
seismic source (or by organized sources not related to the survey)
often satisfy the criterion of a smooth phase function. Acquisition
technology and processing steps have to be designed in such a way
that the final section contains only reflections - and preferably
only primary reflections - while all other events (and eventlike
noise) are suppressed.

Reflections

A plane wave that falls vertically on a plane interface
(between two media) is reflected with amplitude

$$ p_r = p_i \ \frac{z_2 - z_1}{z_2 + z_1} \ = p_i \cdot R $$

where p_i is the amplitude of the incidence wave and Z is the
'seismic impedance', the products of seismic velocity and density
of the two media respectively. A simple estimate shows that the
reflection coefficient R can easily be of the order 0,1 for reasonable assumptions about different rocks near the base of the crust:

$V_1 = 6,5$ km/s $\rho_1 = 2,73$ g/cm^{-3} $Z_1 = 17,745$

$V_2 = 7,1$ km/s $\rho_2 = 3,05$ g/cm^{-3} $Z_2 = 21,655$

$R \approx 0,10.$

Such reflections are easily detected with current acquisition
technology. However, neither the wave fronts nor the interfaces we
are concerned with are plane. One can estimate the effect of wavefront - and interface-curvature with the help of the Fresnel-zone:
one looks at all seismic paths from source to receiver via the
interface in question that deviates from the 'actual ray' - the
path of stationary time in the sense of Fermat's principle - by
less than half a wave length, since energy travelling along any of
these paths interferes constructively with the energy returned via
the 'nominal ray path'. The Fresnel-zone is that area on the interface that is touched by these selected paths. The relative amplitude of a reflection depends on the size of the Fresnel-zone.

Generally, one estimates Fresnel-zones for plane interfaces
and for coincident transmitter and receiver. A rule of thumb for

the radius of the Fresnel-zone in the far field is the following:
if n is the number of wave lengths for the length of the path (or
the product of frequency and reflection time), then one can obtain
the radius r of the Fresnel-zone by multiplying half the wave length
with the square root of n.

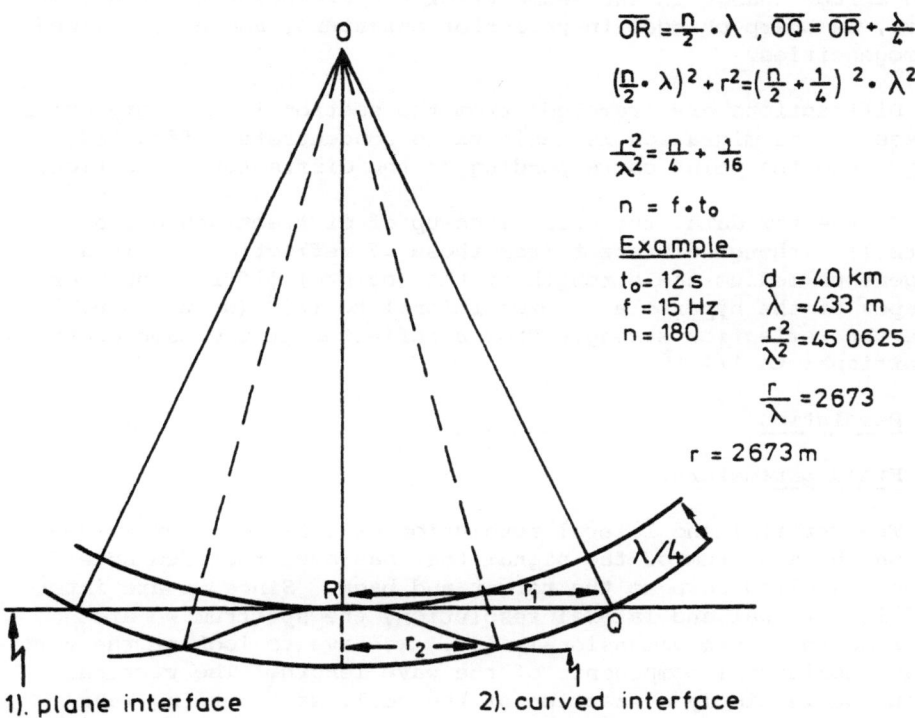

$$\overline{OR} = \frac{n}{2} \cdot \lambda \;,\; \overline{OQ} = \overline{OR} + \frac{\lambda}{4}$$

$$\left(\frac{n}{2} \cdot \lambda\right)^2 + r^2 = \left(\frac{n}{2} + \frac{1}{4}\right)^2 \cdot \lambda^2$$

$$\frac{r^2}{\lambda^2} = \frac{n}{4} + \frac{1}{16}$$

$$n = f \cdot t_0$$

Example

$t_0 = 12\,s \qquad d = 40\,km$
$f = 15\,Hz \qquad \lambda = 433\,m$
$n = 180 \qquad \frac{r^2}{\lambda^2} = 45\,0625$

$$\frac{r}{\lambda} = 2673$$

$$r = 2673\,m$$

1). plane interface   2). curved interface

Fig. 1. Fresnel-zone for a plane interface (1) with the
radius r_1 and for a curved interface (2) with the
radius r_2.

Refractions

Refractions are generally observed only with rather large
offsets, so that refraction events in standard reflection surveys
are rare. However, in crustal studies large offsets are the rule,
so that refraction events from intermediate high-velocity layers
might interfere with the reflection events. Data processing designed to enhance primary reflections (primarily CDP-stacking)
strongly discriminates against refracted events.

Diffractions

Diffractions play a two-fold role in reflection seismics: on the one hand they can be regarded conceptionally as the most basic elements of a reflection section (in the sense of Huygens' principle - this aspect is discussed in some detail in a later section); on the other hand diffractions occur in stacked sections - since the stacking process only weakly discriminates against them - wherever there is an abrupt change in the geometry of the Fresnel-zone, i.e., at faults, at sharp changes in reflector curvature, and at localized inhomogeneities.

Diffractions are 'removed' from the section in the 'migration' process, since migration is designed to concentrate diffracted energy into the point corresponding to the diffractor's position.

In the raw data, the phase line-up of diffractions are a hyperbolae (though different from those of reflections). In a homogeneous medium the strength of the (point-) diffraction near the apex of the hyperbola is proportional to $1/d^4$ (d= the depth) while the reflection strength from a reflector at the same depth is proportional to $1/4\,d^2$.

3. Resolution

Field parameters.

The vertical and lateral resolution of a seismic survey depend on the spectrum of the signal that has been recorded after having travelled down to the target and back. Since we are interested in vertical and lateral resolution, the spectrum we are referring to is two-dimensional, i.e., we have to look at the vertical and horizontal components of the wave length. The vertical resolution is simply a fraction of the smallest occurring wave length (depending on definition betweeen 1/4 and 1/6 of λ). The horizontal resolution depends on the smallest occurring apparent wave length $\lambda_a = \lambda_{min}/\sin\varphi_{max}$ along the surface. If acquisition and processing techniques permit the proper recording of rays with an inclination of 45°, the smallest apparent wave length is 1.414 times the actual smallest wave length. In order to realize this resolution the wave field must be sampled in space and time in such a way that complete reconstruction without loss or aliasing is possible. The sampling theorem guarantees 'no loss, damage, or aliasing' if there are at least two (time-) samples per shortest period and at least two space samples (geophone groups)per shortest apparent wave length. Prudence advises to have at least 2.5 samples per shortest period/shortest apparent wave length.

Realistic figures for a deep crustal survey are fmax = 40 Hz

ms, λ min = 150 m, and λmin, app. = 210 m. Maximum sampling intervals are thus Δt <10 ms and Δx <80 m.

Undersampling in time leads to spurious frequencies in the pass band that can in no way be removed. Spatial undersampling leads to spurious dips. However, a significant difference exists between the two types of undersampling: spatially undersampled data (for instance as consequence of dips larger than expected) can be salvaged by post-acquisition limitation of high frequencies. This results in larger apparent wave length and thus in data that are (spatially) sampled correctly. One has, of course, to pay for this by reduced resolution both in time and space.

Factors that control signal spectrum and amplitude range.

The signal spectrum limits the resolution of a survey: faulty data acquisition can result in lower resolution or spurious correlation, but oversampling in time cannot increase the resolution beyond this limit. It is therefore important to realize what determines the spectrum of the recorded signal.

The signal spectrum is controlled (i) by the spectral transmission properties of the medium between surface and target (and for complicated targets by its spectral reflection properties), and (ii) by the spectral properties of the elements of the acquisition chain: the source spectrum is the maximally available spectrum, but it is further limited by source-and receiver-array characteristics, transmission properties of cables and geophones, and filter settings.

1.   Noise

Closely related to the question of the useful spectrum is that of the useful amplitude range. In abstract theory one can recover infinitely small amplitudes, but in the real world of electronic instruments one is limited on one side by the maximum amplitude geophones and amplifiers can accept without distortion, and on the other side by the thermal noise in the input circuit that is amplified together with the signal. In the field one is further limited by extraneous noise. Part of this noise is of our own making: waves generaged by the source that are not useful to the survey. Since the source generated noise is deterministic, methods can be designed to decrease its influence (e.g., source and geophone patterns, buried shots). The remaining noise is ambient - due to distant earthquakes, surf action, wind motion transfered to the ground by the roots of trees, traffic, industrial machinery, and the like. Noise of the second type - commonly called statistical or uncorrelated - can only be suppressed by statistical methods. These methods are severely limited by the fact that even under the best of circumstances the improvement goes but with the square root of the number

of independent observations: to improve the ratio of the signal
to the statistical noise by a factor of ten, one would have to use
one hundred independent observations. Thus there is a practical
limit to statistical improvement.

Noise has two aspects, amplitude and phase. With some
simplifications we can say that the amplitude noise masks the
phase line up we want to see, and that the phase noise destroys
the line up.

Amplitude noise is partly source generated and partly e -
traneous. Extraneous noise can be suppressed by statistical means:
since its space-time-dependence is different from that of the source
generated signal (and noise), stacking of data obtained at different
times favours the systematic parts of the record.
Source generated noise has to be suppressed by acquisition geometry:

Different paths	directivity recording
Different frequency	frequency depending recording
Different wavenumber	wavenumber depending recording
Different polarization	polarization depending recording.

2.   Instrumentation

The ideal situation is where the electronic instrument noise
is just somewhat lower than the ambient noise, since then the
entire dynamic range of the instrumentation can be used. The
technology of reflection seismics has been dominated by the
question of useful amplitude range.

The development began with the recording of several·shots with
different charges. For a long time the standard of industry was
automatic gain control (AGC). With AGC the useful amplitude range
could be made self-adapting (limited only by the instrument noise
on the low end, and by the maximum excursion of geophone coils, as
well as the maximum compression of input voltages the design
allowed). A severe shortcoming of AGC is the difficulty to recover true amplitudes.

Currently, three types of seismic recording equipment are in
use. All three permit - in principle - recovery of the true geophone output. Well designed geophones are believed to operate
distortion-free over the amplitude range the different systems
cover. While this might be true for well maintained geophones,
it is not necessarily true for geophones that have been in operation for some time without checkup.

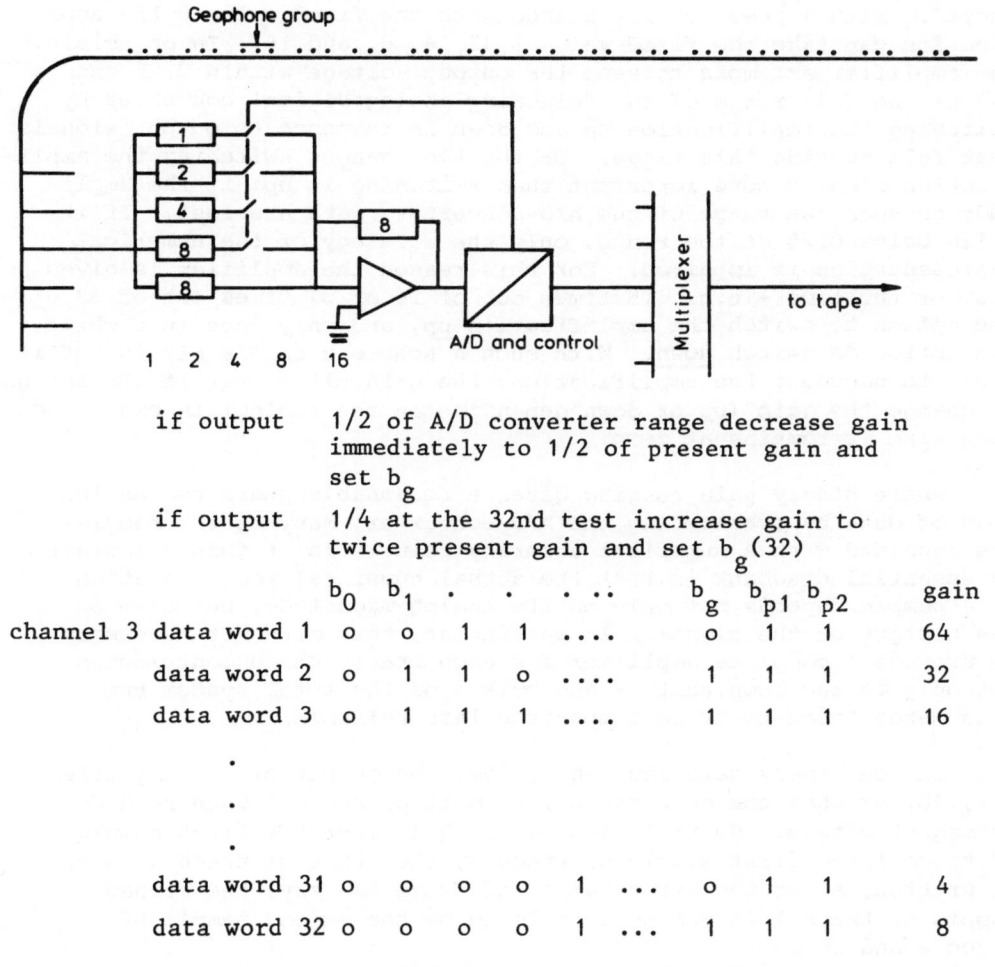

if output 1/2 of A/D converter range decrease gain immediately to 1/2 of present gain and set b_g

if output 1/4 at the 32nd test increase gain to twice present gain and set b_g(32)

	b_0	b_1	.	.	. ..	b_g	b_{p1}	b_{p2}	gain
channel 3 data word 1	o	o	1	1	....	o	1	1	64
data word 2	o	1	1	o	....	1	1	1	32
data word 3	o	1	1	1	....	1	1	1	16
.									
.									
.									
data word 31	o	o	o	o	1	...	o	1	1	4
data word 32	o	o	o	o	1	...	1	1	1	8

Fig. 2.   Binary gain amplifier system.

Binary gain ranging:

Fig. 2 shows the principle of <u>one channel</u> of a 'binary gain amplifier system', representing the technology of about 1965 but still in use.

In the simplified example the third geophone group is connected to an operational amplifier. The input resistance can take the values 8, 4, 2, 1, and 0.5 depending on which of the switches are closed. With a feedback resistance with the fixed value 8 the amplification can take the fixed value 1, 2, 4, 8, and 16. In principle, the amplifier attempts to keep the output voltage within 0.25 and 0.5 of the full range of the following analog/digital converter by switching the amplification up and down in response to output signals that fall outside this range. Of the two changes switching the amplification down is more important than switching it up: if the amplitude exceeds the range of the A/D-converter, data are lost. If it falls below 0.25 of the range, only the accuracy of the numerical representation is impaired. For this reason the amplifier is given most of the time - e.g., 15 times out of 16 or 31 times out of 32 - the option to switch the amplification <u>up</u>, and only once in a while the option do switch <u>down</u>. With such a scheme a single bit is sufficient to document the amplification: the gain bit is set if the option to change the gain (up or down depending on the timing) is exercised, otherwise it remains at zero.

While binary gain ranging gives a reasonable guarantee against loss of data by overloading the A/D-converter, many of the samples are recorded with a numerical accuracy that is lower than necessary. An essential drawback is that the actual numerical representation of a sample depends not only on its analog magnitude, but also on the history of the signal. In particular, this makes it necessary to provide a complete amplifier for each trace, which contributes not only to the complexity - and bulk - of the total system but also makes trace-to-trace comparison less reliable.

In the binary gain ranging system, the output of all amplifiers (24, 48, or what the case may be) is multiplexed and then recorded on magnetic tape. Multiplexing means that after the first sample of trace 1, the first sample of trace 2, then that of trace 3, etc. is written; after the first sample of trace 24, say, the second sample of trace 1 is written, followed by the second sample of trace 2 and so on.

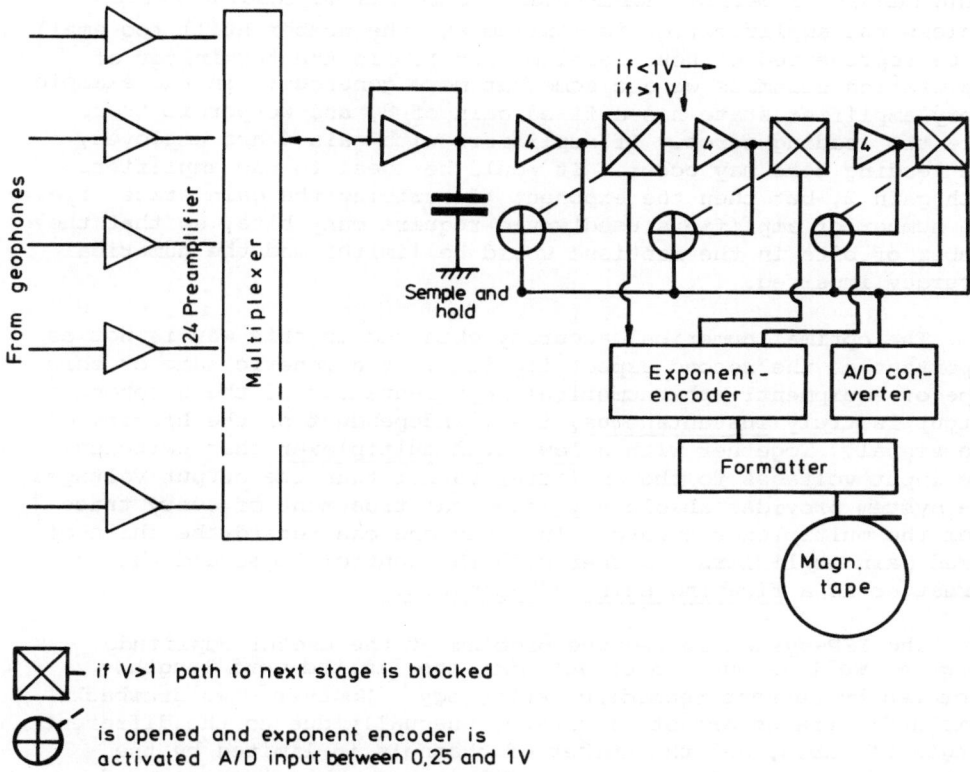

if<1V ⟶
if>1V ↓

Semple and hold

Exponent-encoder

A/D con-verter

Formatter

Magn. tape

From geophones

24 Preamplifier

Multiplexer

⊠ — if V>1 path to next stage is blocked

⊕ — is opened and exponent encoder is activated A/D input between 0,25 and 1V

Fig. 3. Instantaneous floating point amplifier system.

Instantaneous Floating Point Recording:

Fig. 3 shows a simplified Instantaneous Floating Point system representing the state of the art of the seventies and still the standard equipment. The key difference against the predecessor is that the IFP system converts each sample into a <u>normalized floating</u>

Point number: A floating point number consists of an exponent (representing in our case the gain state) and a mantissa (in our case the digital representation of the output of the analog amplifier). It is called normalized if the amplification is as high as possible without loss of data. In computer terminology a floating point number is called 'normalized' if it has no leading zeros (unless the amplification is maximum and the number still too small to be represented without leading zeros). In the terminlogy of exploration seismics we are somewhat more generous: in the example every amplifier state has a fixed gain of 8, and we permit therefore two leading zeros. If amplifiers with gain 4 are employed, one leading zero may occur. It would be ideal to use amplifiers with gain 2, but then the exponent (describing the gain state, i.e. the number of amplifiers used)would require many bits, so that the number of bits in the mantissa would be limited and the numerical accuracy impaired.

The optimal numerical accuracy obtained in this way is not as important as the second aspect implied in the generic name or this type of equipment: the numerical representation of the geophone output is truly instantaneous, i.e., independent of the history of the signal. Together with a low level multiplexer that switches the input voltages to the amplifier rather than the output voltages, the system provides absolutely identical treatment of every trace from the multiplexer onward. In a way one can regard the chain of fixed-gain amplifiers together with the control logic and the formatter as a floating point A/D-converter.

The IFP-system solved the problem of the useful amplitude range as well as one can expect with the limited word length dictated by current recording technology. However, two drawbacks remained: traces are still treated unequally due to the different length of cable, and the number of channels is limited by the number of pairs of leads one can fit into a cable. The second limitation· is more severe at sea than on land, where a second or even a third cable could be laid out.

Telemetry principles (see fig. 4):

At this moment, the most modern seismic equipment is of the telemetry type. In such devices, signal transport from the recording station is done in digital form. This means that every geophone group (streamer section) has its own sample and-hold circuit and its own (floating point) A/D converter. The devices are synchronized with the recording unit. 'Polling' of stations follows a protocol in such a way that the data words from the stations form a multiplexed data stream. Only two pairs of wires are needed: one for the data stream and one for synchronization and polling signals. In areal surveys and in split spread arrangements one uses several datastreams and synchronization lines. These are not directly connected to the formatter-recorder system, but enter a special device (e.g., a Line Extension Module (Sercel)) that temporarily stores data and schedules the transmission to the recorder.

The key advantages of telemetry are the freedom in the choice of channels and the homogeneity of the data. Now every trace is truly identical, since for digital data signal changes on the transmission path or interaction of the transmitting medium with the signal source - at least in principle - do not exist: either the transmission is adequate and the original pulse sequence re-presenting the data can be restored, or it is not, in which case there are no data. (To make sure that the pulses can be restored, cable telemtry uses repeater stations that overcome transmission losses and re-shape the pulses).

On land and at sea the number of channels is, for a non-telemetric system, limited by the number of wire pairs in cables and streamers. With telemetric systems the only limited factor is the number of words that can be put into a data stream without overlap during one sample time.

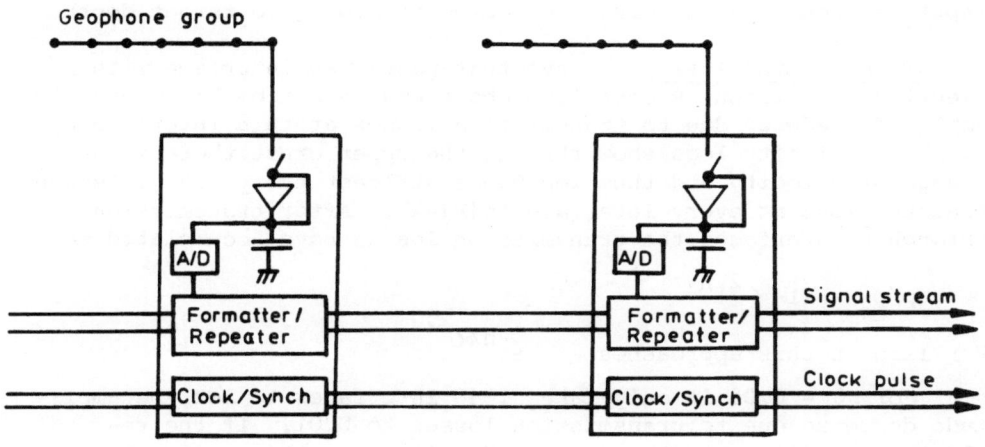

Fig. 4. Telemetric field stations.

3. Propagation effects

It is suggested that the reduction in amplitudes caused by spherical spreading, absorption, and transmission losses scale in such a way that there are no significant differences between 'deep' and 'shallow' reflection surveys (barring the operational difficulties of scaling).

Spherical spreading: In a homogeneous subsurface, the amplitude of seismic waves decreases since the energy is spreading over a larger and larger surface. Since, on one hand, the surface of a sphere is proportional to the square of the radius and, on the other hand, the energy contained in a seismic wave is proportional to the square of the (pressure) amplitude, the amplitude at distances r_1 and r_2 follow the simple rule

$$p(r_1) \cdot r_1 = p(r_2) \cdot r_2 .$$

From this it would follow that the amplitude decay in a deep survey is much larger than in a shallow survey. However, for explosive generation of seismic waves the signal amplitude, as well as wave length and the equivalent cavity are proportional to the third root of the charge. If one denotes the pressure at the equivalent cavity radius with p_o and the cavity radius with r_o, one finds

$$\frac{p}{p_o} = \frac{r}{r_o} = \frac{r}{\lambda} .$$

The amplitude decay due to spherical spreading is properly scaled with the dominant wave length. One should expect that deep surveys are as difficult as shallow ones, provided total energy input and source array size are scaled according to target depth.

Transmission losses: A wave that passes an interface with reflection coefficient R once from above and once from below has its amplitude reduced due to transmission losses at this interface by $(1-R^2)$. Velocity logs show that in the upper crust there is a change of velocity and thus impedance at least every 0.3 m. Assume that the loss at every interface is $|R|=\alpha$. After transmission through n interfaces the transmission losses have accumulated to

$$p/p_o = (1-\alpha^2)^n .$$

For large n this approaches $e^{-n\alpha^2}$

For α = 0.05 it only takes 1840 interfaces to let the amplitude decrease due to transmission losses to 0.01. If the reflection coefficient α is only 0.01, it takes the passage through

146

about 4600 interfaces to reduce the amplitude to 0.01. With about 3 interfaces per m this would require only about 1.5 km for a sizable amplitude reduction. From this it would appear that there should be no reflections from any larger depth unless the upper crust is extraordinarily - and uncharacteristically - homogeneous. What is wrong with this reasoning ?

The first argument why such exorbitant losses are not observed was given by O'Doherty and Anstey (1971): if the layering is cyclic, short range multiples (that have traversed only one thin layer and are then following the primary) are nearly in phase with the original signal (fig. 5). Thus much of the energy lost in transmission is added again to the signal. Again it depends on the signal frequency what kind of time delay can be regarded as sufficiently small.

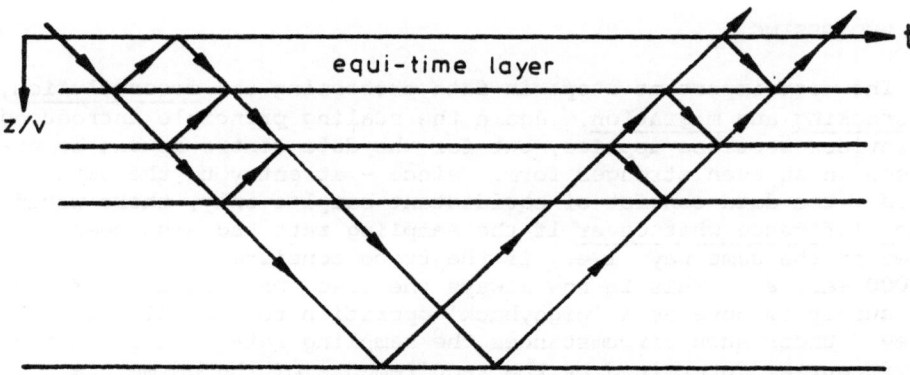

Fig. 5. Short range multiples in a cyclically layered medium. The two reflection coefficients for downgoing waves at the top and bottom of every layer are equal in magnitude $|R|$ and opposite in algebraic sign. Since one reflection is from above and one from below, the compound reflection coefficient, i.e., the product of the reflection coefficients at top and bottom - of all layers is $+R^2$, and the time delay corresponds to $2\,\Delta z/v$.

A correct treatment of the situtation would be to solve the wave equation in thinly layered media. A simple calculation (Helbig 1981a) for periodically layered media and for SH-waves has shown that layers below a thickness of $\lambda /2$ are not 'seen' by the wave if they form part of a repeating stack, but that such a stack is equivalent to a homogeneous - though transversely anisotropic - medium. From these calculations it follows that for transmission losses the wave length again is the proper scaling factor, and that the only significant dimensionless quantity is the number of wave lengths along the travel path (or the number of periods for the reflection time).

Absorption: Most observations of actual absorptions (i.e.,of internal losses of energy in a homogeneous medium) as well as most of the suggested mechanisms for absorption lead to a constant loss of energy per wavelength (or cycle). Thus again we find that - with equal absorption coefficients - the amplitude loss due to absorption should depend only on the number of wavelengths over the total travel path.

The above means that the three main effects responsible for the decrease of amplitudes scale with the wave length. If all things would be equal (and source energy and source array size would be scaled accordingly) deep surveys should not differ significantly from shallow ones.

DATA PROCESSING

The most important steps in data processing are deconvolution, CDP stacking and migration. Again the scaling principle introduced in previous sections applies, but for the data centre it can be expressed in an even stronger form: since - at entry to the data centre - the data consist of equidistant samples only, there would be no difference whatsoever if the sampling rate had also been scaled in the same way, i.e., if the trace consists of about 1000 to 2000 samples. This is now always the case, particularly if the deep survey is done as a 'piggyback' operation to a shallower survey. Under such circumstances the sampling rate at acquisition is, of course, determined by the requirements of the commercial survey (see, e.g., the example in fig. 13). Even then a simple resampling operation (in many cases simply the dropping of every other sample) would bring the data into this form.

Deconvolution is typically a single trace operation and will not be discussed further. CDP-stacking and migration are the two standard multi-channel processes. Since a proper understanding of their operation - and their interaction - is important for any kind of modern seismic survey, they are discussed in some detail.

1. Migration:

Migration is a data process that, in principle, is aimed at the determination of the structure that has given rise to the observed wave field. For details see Helbig (1981b) and Hosken (1981).

Migration before stack: Conceptually simplest is the discussion of the so-called 'migration before stack', though at present migration after stack is certainly closer to being a standard process. Imagine that the data volume consists of S times R traces identified by source- and receiver numbers (fig.6, left). Imagine further an originally blank 'section' in which we want to enter the information of whether a point is a 'scatterer' (a source of secondary waves in the sense of Huygen's principle). Such a point is indicated on the right side of the figure with P. If this is a source of secondary waves, these secondary waves should have left their mark on essentially all traces, and the simplest way to ascertain the presence of these marks is to sum the appropriate instants from all traces. The appropriate instants on the traces can be found by observing that the time should be equal to that required by the wave to travel from the selected source to the (presumed) scatterer and on to the selected geophone. If the point was a scatterer, the signal amplitude at the times so determined should have the same algebraic sign on all traces (barring noise and interferences), and the sum has a large (positive or negative) value. Noise and interferences are, of course, present in real data, but since they are not organized in the same way as primary reflections, all other events (and all random noise) are discriminated against by the stacking process.

We have talked here not of primary reflections, but of 'scatterers', i.e., of diffractions. However, a reflector can be regarded as a closely spaced set of scatterers (and the reflections as the superposition of the secondary waves generated by these scatterers), thus the argument holds for reflections in the same way as for diffractions. The discrimination against multiple reflections is based on a subtel argument: in defining the correct instants on the different traces one has to make use of the correct velocity distribution. Since velocity increases with depth, and since multiples have travelled in shallower regions than primaries with the same travel time and thus have experienced a lower average velocity, a set of instants (a phase function) that would be correct for primaries cannot be correct for multiples. This argument holds even more strongly for any refracted event that is contained in the data set: since a major part of the refracted ray path is nearly horizontal, the average velocity experienced by refracted arrivals must be significantly lower than that of primary reflections with the same travel time.

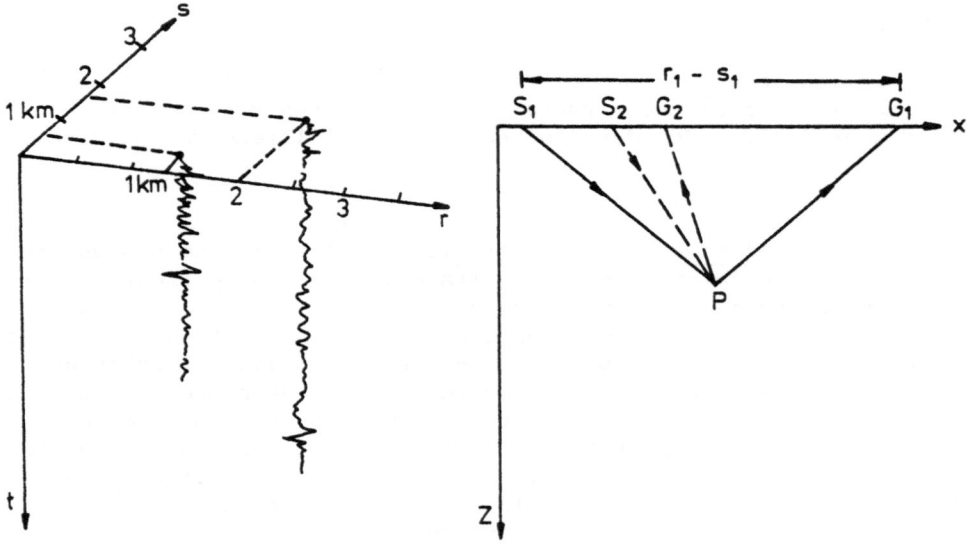

Fig. 6. Principle of 'migration before stack'. Left: two of
the traces that make up the 'raw data volume',
identified by source position s and receiver position
r. Right: actual position of sources. If the point
P has scattered the seismic wave (i.e., if it is a
source of secondary waves in the sense of Huygens's
principle), the corresponding signals are to be
found on the traces and times corresponding to the
travel times along the ray paths S_1PG_1 and S_2PG_2,
respectively.

Migration after stack: The 'raw data volume' depicted at the
left hand side of figure 6 consists of (R) times (S) traces, where
(R) is the number of recording channels and (S) is the number of
separate source groups taken into account in the migration process.
Both (S) and (R) have orders of magnitude of several hundred, the
number of samples per trace is of the order of 1 000 to 2 000, so
the raw data volume involved in a migration-before-stack process
generally consists of well above ten million data words. In the
process, about one thousand potential scatterers have to be investigated for every original receiver position (i.e., with group
separation 100 m there would be 10 000 for every kilometer of line),

and each of this 'investigations' involves the summation of about 10 000 individual data values (which, moreover, generally have to be interpolated between the samples). It is thus not surprising that migration before stack is not often applied.

2. <u>CDP stacking</u>: To reduce the amount of data involved in the number of operations required, one compresses the raw data volume by Common Data Point stacking. To this end one re-orders the traces into 'common data point gathers' (the 'Data Point' of the trace is the point midway between source and receiver on the datum plane) and then 'stacks' the data of one gather into a simulated zero-offset trace with source and receiver both at the data point. The phase function along which this stacking (a simple or weighted summation) has to take place is a hyperbola if the medium is homogeneous, and does not deviate too much from a hyperbola if the medium is (moderately) vertically inhomogenous. To show this we discuss the reflection from a dipping interface below a homogeneous layer. In the re-ordered data set the data point has coordinate x, and the distance from data point to either source or receiver (the 'half offset') is denoted by y: $x = (s + r)/2$, $y = (s - r)/2$ (see figure 7). The 'target path' to which we want to transform our data is indicated by h. The path length of the actual ray from S to R is identical to the distance SR'. From the law of cosines follows

$$\overline{SR'}^2 = 4h^2 + 4y^2 + 4y^2 \cos^2 \alpha$$

and thus with $t(x,0) = 2h/v$

$$t^2(x,y) = t^2(x,0) + (2y)^2 \frac{\cos^2\alpha}{v^2}.$$

If $v/\cos$ is known, we can calculate with this equation the time shift that is necessary to transform a non-zero offset trace into a zero offset trace:

$$t(x,y) = t(x,0) + \Delta t$$
$$t^2(x,y) = t^2(x,0) + 2t(x,0)\Delta t + \Delta t^2$$
$$\Delta t^2 + 2t_o \Delta t - (2y)^2 \cos^2\alpha / v^2 = 0$$
$$\Delta t = t_o(1 + \sqrt{1 + (2y)^2 \cos^2\alpha / (vt_o)^2})$$
$$\approx 2y^2 \cdot \cos^2\alpha / (vt_o)^2.$$

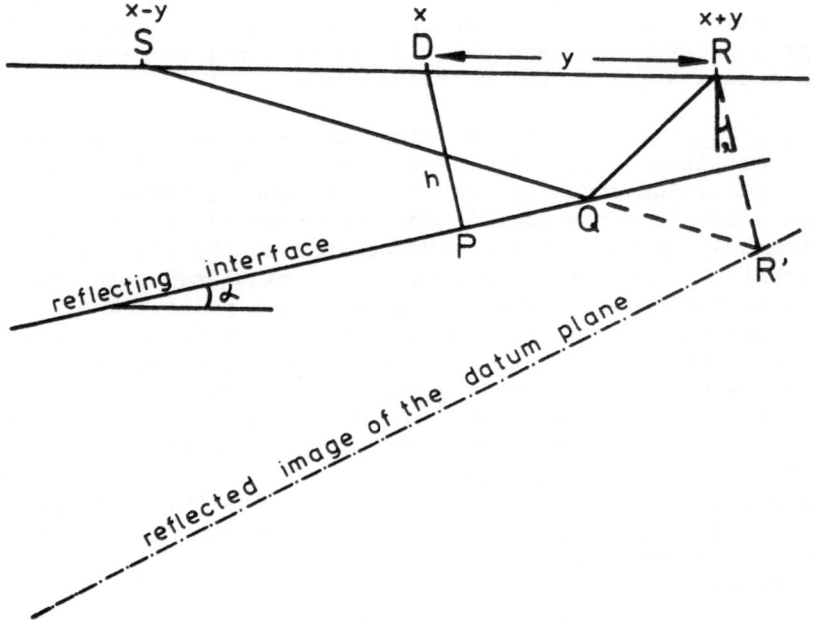

Fig. 7. Reduction of non-zero-offset observations to zero-offset observations (S = source, D = data point, R = receiver).

Velocity estimation

Generally, $v/\cos$ is not known beforehand, but has to be determined from the data. One way to do this is to plot t^2 versus y^2 (see fig.8). If the medium is homogeneous the data follow a straight line with intercept $t^2(x,0)$ and slope $\cos^2\alpha /v^2$. Instead of this intermediate plotting step one can stack the data experimentally with different values of $\cos^2\alpha /v^2$ and accept that value that gives the highest amplitude in the stacked trace.

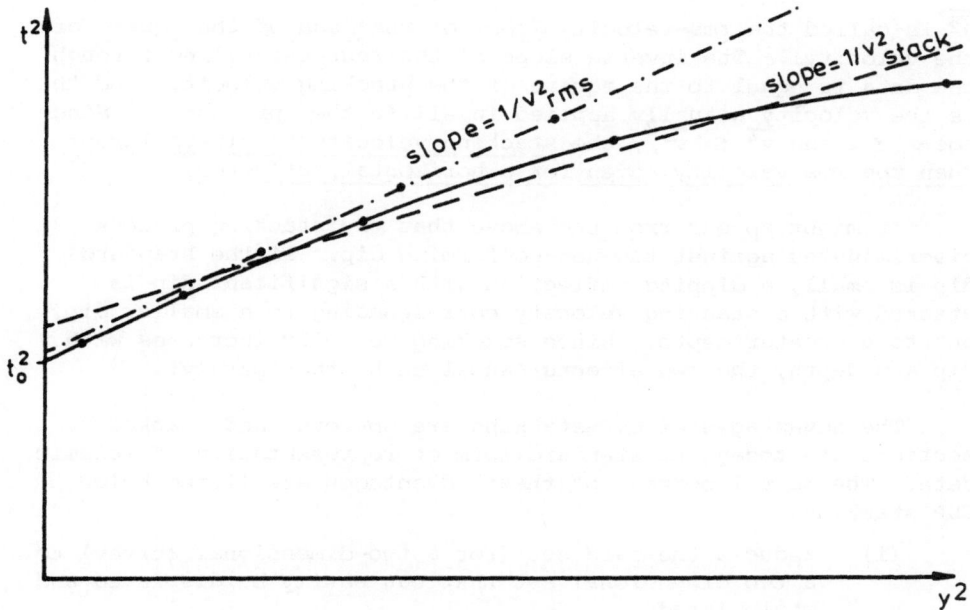

Fig.8. t^2 versus y^2 plot for a vertically inhomogeneous medium.

In a vertically inhomogeneous medium the above arguments have only slightly to be modified: instead of falling on a straight line the t^2 vs. y^2 points fall on a curve with intercept $t^2(x,0)$. For a horizontal reflector ($\alpha = 0$) the first terms of the Taylor expansions are

$$t^2(x,y) = t^2(x,0) + (2y)^2\ \overline{v^2} - 4y^4\ \frac{\overline{v^4} - \overline{v^2}^2}{4t^2(x,0)\cdot\overline{v^2}^4}$$

where

$$\overline{v^n} = \int_0^{t_0/2} v^n dt\ /\ (t_0/2).$$

$\sqrt{\overline{v^2}}$ is called the rms-velocity (root of the mean of the square of the velocity). The inverse slope of the regression line through the data is equal to the square of the stacking velocity, and this is the velocity actually applied in all further processes. Since $\cos\alpha \leq 1$ and $\overline{v^4} \leq \overline{v^2}^2$, the <u>stacking velocity is always larger than the rms velocity, even for a horizontal reflector</u>.

It might appear from the above that the stacking process discriminates against the non-conforming dip. If the standard dip is small, a dipping reflection with a significant dip is stacked with a stacking velocity corresponding to a smaller dip, but to a greater depth. Since stacking velocity increases with dip and depth, the two effects cancel each other partly.

The advantages of CDP-stacking are obvious, and stacked sections are today the standard form of representation of seismic data. The most important of these advantages are listed below. CDP stacking

    (i)   reduces the data set (for a two-dimensional survey) to a two dimensional set that can easily be displayed and manipulated.

    (ii) improves the signal-to-noise ratio by discriminating against events with phase functions (t-y functions) that differ from the function for the events used in the determination of the stacking velocity. The most important events thus discriminated against are multiple reflections. In surveys in sedimentary layers this is the key argument for the use of CDP-stacking.

    (iii) provides a relatively simple method of velocity determination, even if the velocities thus determined have to be used with caution.

Migration after stack

The stacked section can be regarded as an approximation of a zero-offset section (with all non-primary and non-diffraction events decreased in amplitude). The migration of stacked data is thus - compared with the migration of unstacked data - a relatively simple process (see figure 9): to decide whether a point P is a scatterer, one only has to sum the data along a hyperbola in the x-t plane.

Limitations to migration after stack: From the above discussion one might get the impression that migration after stack gives the same result - at a considerably lower expenditure - as migration before stack. Actually, this is not completely correct. The discrepancy can be traced back to a discrepancy in reasoning in the previous section: in figure 7 we discussed a plane reflector (with different rays of the CDP gather reflected at

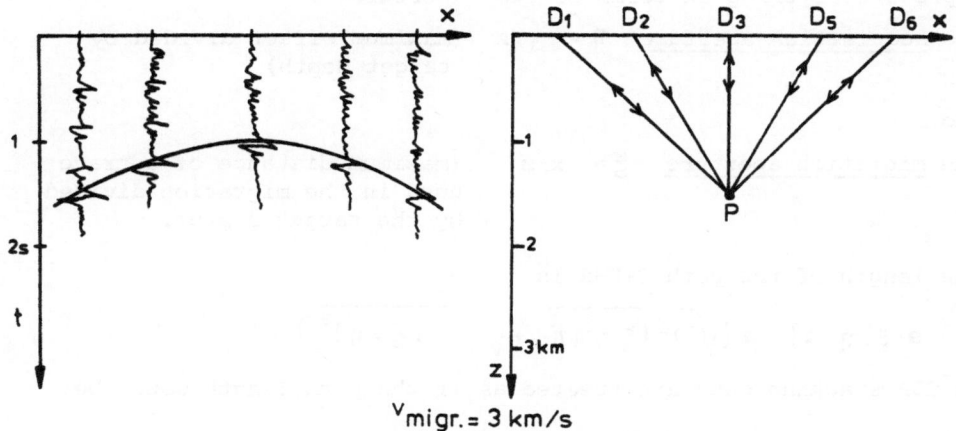

$$V_{migr.} = 3 \text{ km/s}$$

Fig. 9. Migration after stack (P = a scatterer).

different points P and Q, while in figure 9 we treat the stacked
traces as true zero-offset traces from data point D_i to diffractor
P.

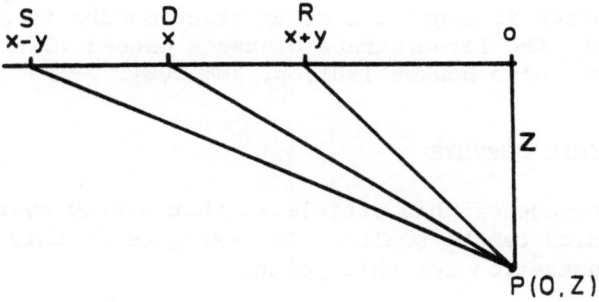

Fig. 10. Ray path of a CDP gather for a single scatterer.

Let us briefly discuss the magnitude of the errors thus induced.
Figure 10 shows a ray path S-P-R belonging to a CDP gather centered
at D. The medium has been assumed to be homogeneous. We shall
express everything in terms of two 'apertures':

the acquisition aperture $\eta = y/z$ (maximum offset divided by
target depth)

and

the migration aperture $\xi = x/z$ (maximum distance of a trace
used in the migration divided
by the target depth).

The length of the path S-P-R is

$$s(\xi, \eta, z) = z \left(\sqrt{1+(\xi-\eta)^2} + \sqrt{1+(\xi+\eta)^2} \right).$$

In CDP stacking data are treated as if the path length would be

$$\bar{s}(\xi, \eta, z) = 2z \sqrt{1+\xi^2+\eta^2} = s(\xi, \eta, z) - \Delta s.$$

Stacking with subsequent migration can be only successful
with $\Delta s < \lambda/2$.

By expanding the above expressions we get

$$\Delta s/(2z) = (1+\xi^2+\eta^2) \left[\sqrt{1 - \frac{1}{2}\left(1 - \sqrt{1-(\frac{2\xi\eta}{1+\xi^2+\eta^2})^2}\right)} \quad -1 \right].$$

If one again assumes that $t_o \cdot f = 2z/\lambda = n$, the condition for
successful migration after stack becomes

$$\Delta s/z = 1/n,$$

i.e., the higher the resolution the more stringent the conditions.
Figure 11 shows the situtation for the worst case ($\xi = \eta$).
This is to be read as follows: if the target is at a depth of
500 λ , degradation is migration after stack occurs if both the
maximum offset and the largest trace distance exceed .3 times the
target depth, (see also Hosken 1981, p. 206-208).

EXAMPLES OF SEISMIC SURVEYS

The main message of this article is that nearly everything in
reflection seismics can be scaled. The examples in this chapter
serve as an illustration for this point.

It is often not possible to distinguish between a shallow
section and a deep section on the basis of appearance. What
differences there are is often more a consequence of different
than of different target depth.

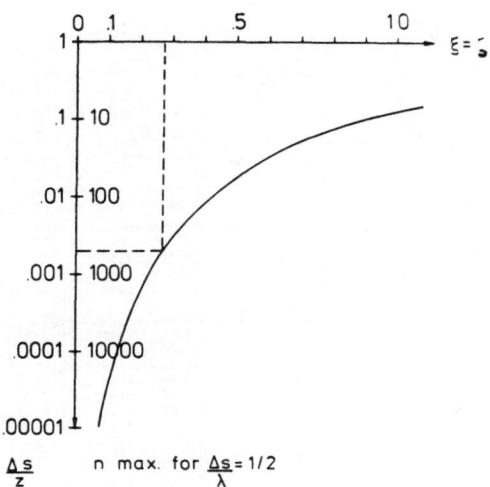

Fig. 11.   Relations between reflector depth, wave length,
           and phase error produced by migration after stack.

Fig. 12 shows a shallow seismic section obtained by students
of the Vening Meinesz Laboratory of the University of Utrecht.
This survey employed single geophones, fixed gain recording,
weight drop as source, 6 fold CDP-stack. Maximum reflector depth
is 200 m. Dominant signal frequency is 200 Hz resulting in high
resolution, especially for the structurally complicated near-
surface layers. Inspite of the relatively low acquisition effort
this section is not much different from one obtained with 40 Hz
signal frequency at depths between 1 and 2 km.

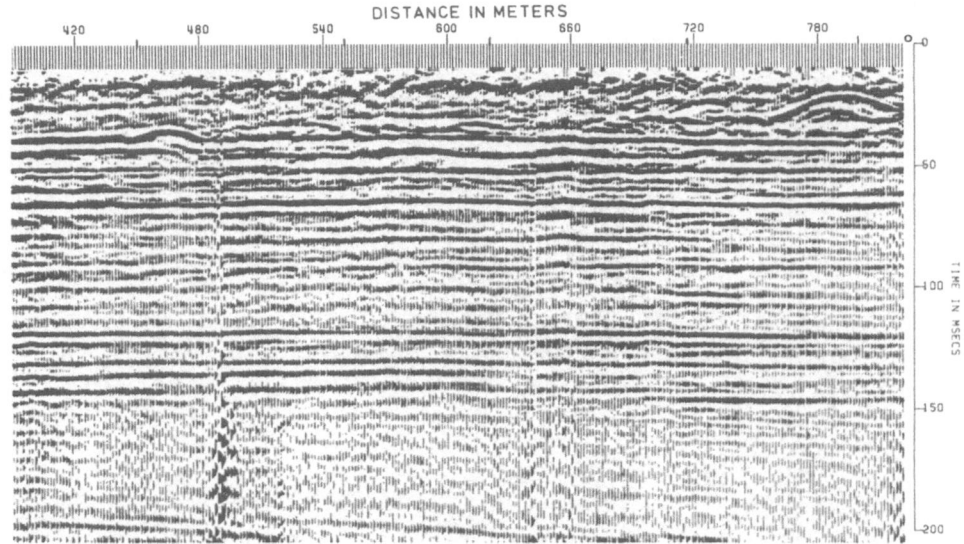

Fig. 12. Shallow seismic line, Plaat van Oude Tonge,
Zeeland, single geophone, fixed gain, weightdrop,
6-fold CDP-stack (students practicum).

Fig. 13 shows the result of a deep reflection survey carried
out in Northern Germany. The source was dynamite and the number
of geophone groups per shot was 48 with a 60 km distance between the
geophone groups (spread length about 3 km).

Multiple coverage was 24-fold. The reflections at the top
of the section (down to about 2 s) come from the Mesozoic and
Zechstein; the good reflections in the range of 6 to 11 s originate from the deep crust. The part of the line shown in Fig.12

158

Fig. 13. Deep reflection line in Northern Germany. Multiple
geophones, instantaneous floating point recording,
dynamite 24-fold CDP stack (commercial survey),
surveyed and processed by PRAKLA-SEISMOS.

has a length of about 30 m. The data acquisition and processing
parameters do not essentially differ from those of normal surveys,
although great efforts were made throughout to optimize the
signal/noise ratio.

   A special seismic field arrangement was chosen for a combined
reflection-refraction survey in the region of the Urach geothermal
anomaly (Meissner et al., 1982, Bartelsen et al., 1982). The objective of this survey was to find out how seismic parameters,
especially horizontal and vertical velocities, are related to the
geothermal anomaly. Laboratory investigations of various materials
under high pressure and high temperature conditions have shown

that a temperature change of 100° at constant pressure gives roughly a 2% change in v_p-velocity (Kern 1978, Meissner et al., 1980). This value is valid for p-T conditions in continental crust. Effects of inhomogeneity and anisotropy may be much stronger, and careful investigations have to be carried out in order to distinguish between these effects. On the other hand, a velocity variation of 1 % can be measured by special arrangements in a seismic field survey. In order to determine such small lateral velocity variations down to depths of 30 km different requirements have to be fulfilled for seismic reflection and refraction work. For reflection work large move-out times are required, e.g. about 1 s for 10 s two-way travel time. These large move-out times were obtained by a continuous spread length of about 23 km, using 144 geophone groups per shot with a group interval of 160 m. The multiple converage was 8-fold.

Because of the large spread length and the considerable variations of the sediments along the lines careful static corrections were important, and therefore additional special stripping corrections after Krey (1978) were applied. The results of refraction work served as a base for the stripping to about 3 km (Fig. 14). The stripping comprised a complete linearization of the isovelocity lines in the sedimentary layers and in the uppermost cristalline. Subsequent to stripping the entire sedimentary region is represented by a comparatively thin layer with a velocity of 4 km/s. At depths larger than 2.5 km the velocities within the crystalline crust are a rough estimate based on the results of refraction seismics (Giese et al., 1976).

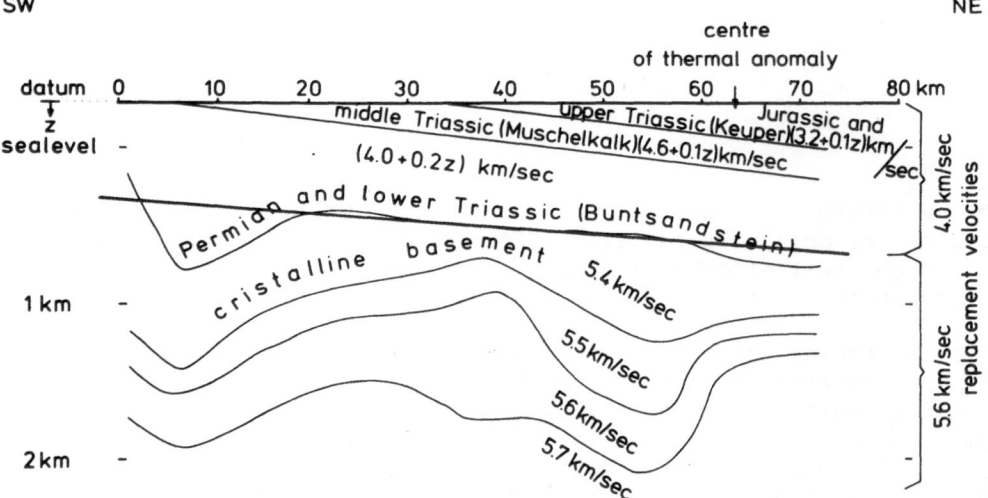

Fig. 14. Velocity model for stripping corrections

centre of geothermal
anomaly

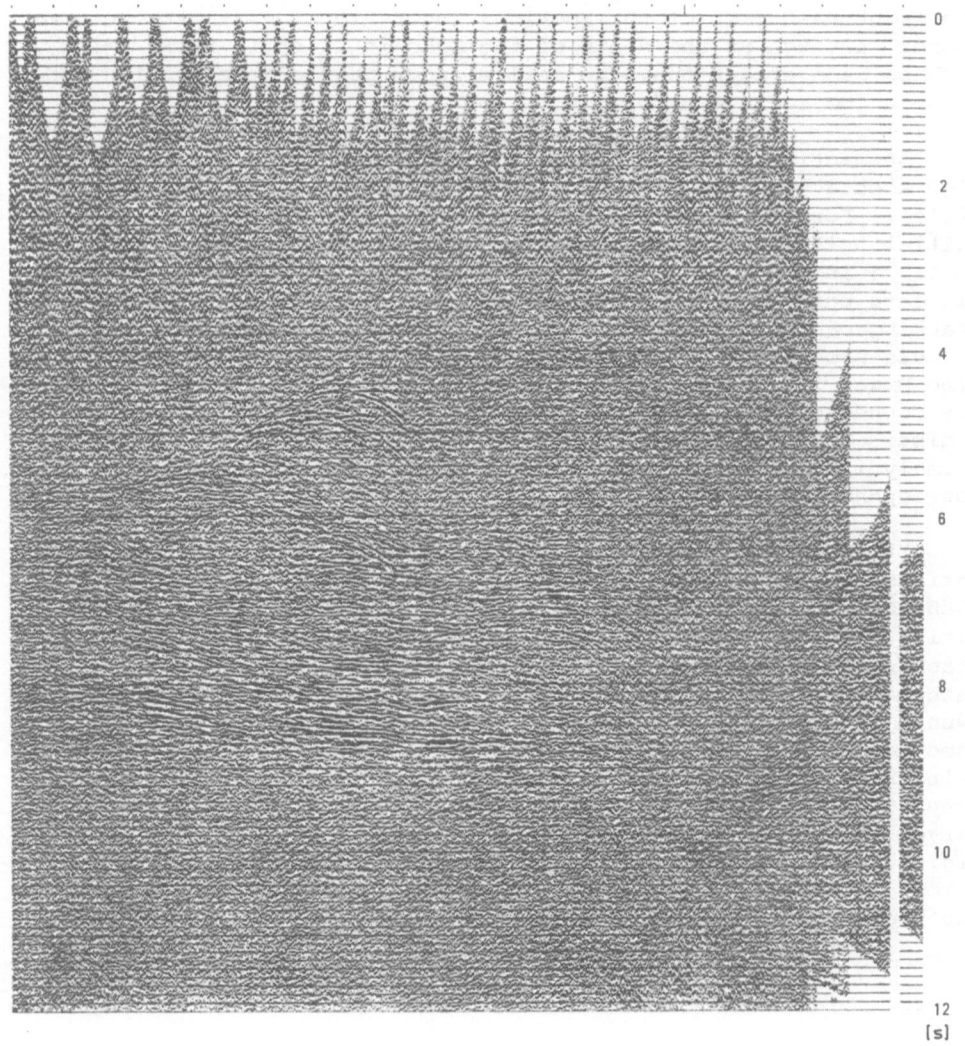

Fig. 15. Stacked section of Urach line 1 (surveyed and
processed by PRAKLA-SEISMOS GmbH.

Fig. 15 shows the stacked section of line 1 (about 80 km) with, on average, a 3-fold vertical exaggeration. The most outstanding reflections originate from the lowermost part of the crust or the Moho (at 8 to 9.5 s). However, there is another group of reflections between 4 and 7 s which indicate unconformities, in particular lenses and layer thickenings from left to right (WSW to ENE). In the range of the Urach geothermal anomaly (on the right hand part of the section) the reflection quality becomes poorer, altough at the depth of the Moho it is still fair. With perhaps same exceptions, the entire upper part of the section seems to be a "no reflection" range.

As mentioned before, the long geophone spreads permitted high accuracy in the determination of stacking and interval velocities. For the determination of optimum stacking velocities an initial uniform velocity model was changed systematically from 91 to 110 % in steps of 1 %. Stacking was thus repeated 20 times for all data points. In most cases the percentages yielding optimum stacking can easily be decided upon. Figure 16 shows the result of this procedure for reflections of two time intervals along the line Urach 1. The points contain root-mean-square error bars and are connected by a smoothed line. For a deeper reflection (b) a distinct low velocity body in the region of the Urach geothermal anomaly is clearly defined. For the shallower reflections a corresponding trough is less clear (a).

It is well known (Krey 1976, 1978, 1980) that the lateral variations of the optimum stacking velocity are not identical with the variations of the average velocity or of interval velocities. The Problem of converting the observed lateral changes of stacking velocities into lateral changes of average velocities was mathematically solved, applying certain simplifying assumptions, by Lynn & Claerbout (1979). Here the problem has been solved in a somewhat rigorous manner by referring to Hubral (1980). Then, using Krey's method (1976, 1980a, Bartelsen et al. 1980, Hubral and Krey 1980), different velocity models and their fit with the observed variations of optimum stacking velocities were computed. Fig. 17 presents the final velocity model after four steps of interation. It shows a large body of reduced velocity (hatched area) in the region of the Urach anomaly.

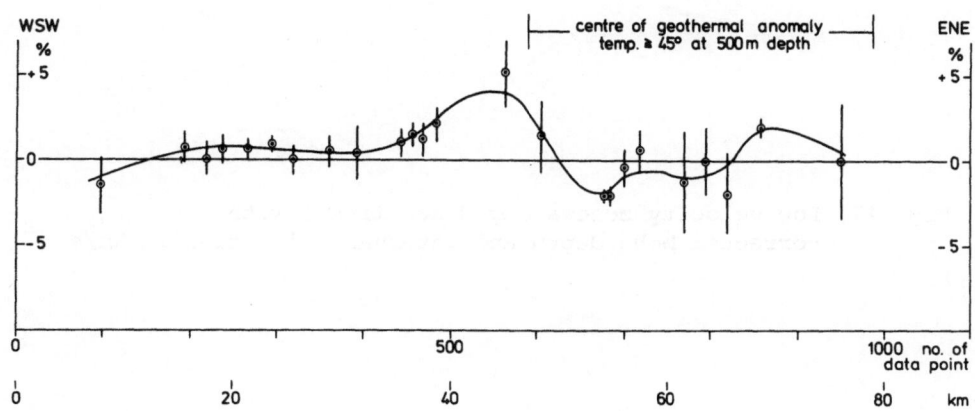

(a) Reflection times between 5 and 7 s

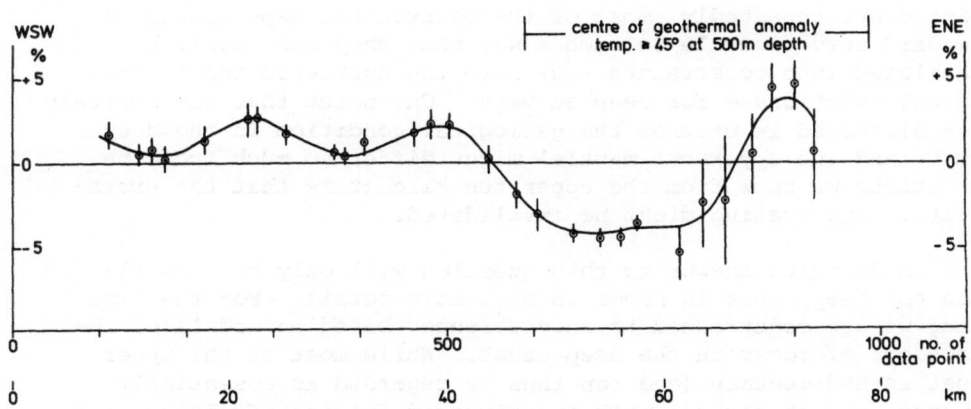

(b) Reflection times between 7.5 and 10s.

Fig. 16. Deviations of observed stacking velocities from
the original velocity assumptions along Urach
line 1 for two time intervals.
Computed points are provided with rms error bars.

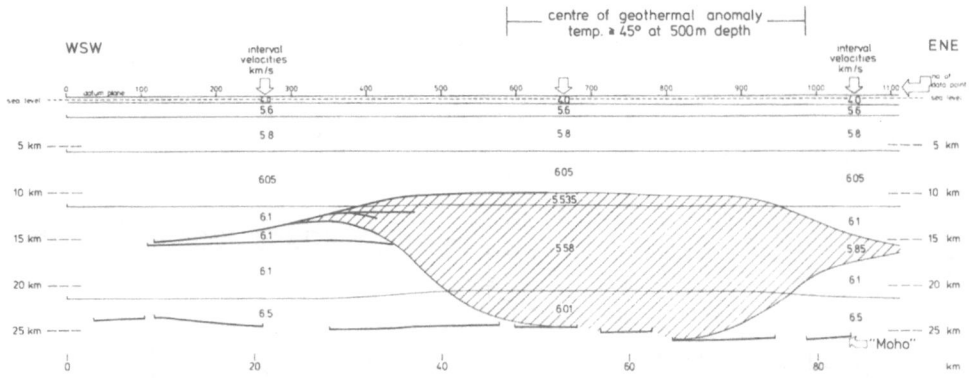

Fig. 17, Low velocity zone along Urach line 1 with
corrected Moho depth and interval velocities in km/s

CONSEQUENCES FOR DEEP CRUSTAL SURVEYS

In the previous chapters we have reviewed the state of the
art of seismic exploration with near-vertical reflections. As
pointed out repeatedly, most of the constraints experienced in
standard surveys scale in such a way that they can easily be
translated into constraints - or into the corresponding technological solutions - for deep surveys. One point that has scarcely
been discussed is whether the geological condition in the deep
crust (and the uppermost mantle) might differ so much from the
conditions we know from the upper ten kilometers that the extra-
polation and scaling might be invalidated.

A definite answer to this question will only be possible
when the deep crust is known in much more detail. For the time
being we are constrained to interferences based on models of im-
placement of rocks in the deep crust. While most of the upper
crust is sedimentary (and can thus be regarded as essentially
layered, though perhaps with superimposed folding, faulting, and
tilting), structures in the lower crust are believed to be due to
intrusions. At first sight there seems to be no particular reason
why the upper boundary of an intrusive body should be nearly horizontal, but it is likely that the sheer size of the intrusions in
question have lead to such large gravitational forces that a sort
of equilibrium along par-horizontal surfaces was established. This
assumption could explain the fact that deep crustal reflections
of near-horizontal attitude are observed without forcing us to
change too much of the picture we have of the deep crust. As a

matter of fact, the situation in the deep crust would, in some respect, be more favourable to the generation for observable reflections: an intrusion would lack the rapid fluctuation of velocities and densities with depth one encounters often in the sedimentary sequence. This would lead to a significant reduction in transmission losses. On the other hand it is likely that the significantly higher amplitudes result in higher absorption, so that the overall effect might cancel.

What then can be learned from the experience accumulated in seismic exploration ? We feel that nearly every step in acquisition, processing, and display technology deserves scrutiny and close comparison with what is either current standard or a serious possibility in commercial exploration. It goes without saying that quality often can not be had without expenditure, and that the financial possiblity in a scientific enterprise might differ markedly from those in the industry. However, there is one kind of improvement that is nearly always possible without exorbitant expenditure: meticulous attention to detail. Obviously, one has to pay for such an effort by reduced production, but it is generally possible to postpone some of a planned survey to a later date. Information that is lost to inaccurate acquisition technology can only be recovered at a subsequent and more careful survey.

Since the frequencies in deep surveys are of necessity lower than in a shallow survey, many of the quality control requirements in acquisition could be relaxed, but again this is essentially only a scaling. This refers to source synchronization (within phase angles of about 20°), definition of source-and receiver locations, particularly if patterns are employed to suppress horizontally travelling high-amplitude events (within about 5/100 of the wave length), and the determination of static corrections to better than 1/10 of the dominant period.

The last point is presumable that where the largest improvement is possible, since it is a typical weak spot. It is quite obviously not sufficient to apply static corrections based on elevations only, since the thickness of the low-velocity layer can change rapidly from one geophone station to the other. Moreover, with the larger station distances employed in deep surveys, there is a deeper zone near the surface about which no information is available from the survey data themselves. To ascertain that subsequent data processing can be carried out in optimal fashion, the time delays produced by changes of thickness of near surface layers down to absolute 4 times the distance between geophone stations should be obtained independently.

Another point that deserves some attention is the size of source-and receiver arrays. There has been for a long time a tendency to use large arrays to assure suppression of surface waves. However, if an array gets so large that the infra-array timeshifts due to near-surface conditions become significant, the method becomes self-defeating. If static corrections between geo-phones of an array become of the order of 1/20 of the dominant period, other methods of surface wave suppression must be found. Fortunately, this is easiest where it is most needed, namely for Vibroseis surveys. Even the strong surface waves of a Vibroseis survey are generally small enough to be recorded without dis-tortions. The solution in this case is to use small groups or individual geophones, but a greater number of acquisition channels. Arrays then can be synthesized <u>after</u> static corrections have been applied. Unfortunately this cannot be done without a complete redesigning of a survey.

Commercial data processing has reached a considerable level, and it is questionable whether that same level can be attained in a scientific institution. Of cource, data processing can be re-peated at somewhat lower cost than a new acquisition campaign, and - from a philosophical standpoint - data processing does not add to data, but can only reduce the total information. Nevertheless it is important to realize that the postponement of an important data processing step quite often has consequences beyond a mere delay. Conclusions, planning of the next survey and other things are based on the display that is available, and the incentive to apply a new process to old data is generally not very high.

ACKNOWLEDGEMENTS

The authors are grateful to PRAKLA-SEISMOS GMBH for the support in the preparation of this paper. The seismic survey in the region of Urach was carried out under the direction of Prof. Meissner (University of Kiel) with the support of the Bundes-ministerium für Forschung und Technologie (Federal Republic of Germany), and of the European Community. The authors thank the oil companies PREUSSAG AG, BEB Gewerkschaften Brigitta und Elwerath Betriebsführungsgesellschaft, C. Deilmann AG, Gewerk-schaft Norddeutschland, Itag, Hermann von Rautenkranz, Inter-nationale Tiefbohr GmbH & Co KG, Mobil Oil AG in Deutschland, Deutsche Schachtbau-und Tiefbohrgesellschaft mbH, Deutsche Texaco AG, Wintershall AG. for making the deep reflection section from Northern Germany available.

REFERENCES

Bartelsen, H., Krey, Th., Meissner, R., and Schmoll, J. (1980)
    Information on structure and physical rock properties
    derived from stacking velocities. Paper presented at the
    50th Annual SEG International Meeting, Houston, Texas.

Bartelsen, H., Lueschen, E., Krey, Th., Meissner, R., Schmoll, J.
    and Walter, Ch. (1982), The combined seismic reflection-
    refraction investigation of the Urach geothermal anomaly.
    In: The Urach Geothermal Project, R. Haenel (Ed.),
    Schweizerbart'sche Verlagsbuchhandlung, Stuttgart: p.247-262.

Giese, P., Prodehl, C., and Stein, A., (1976). Explosion Seismology
    in Central Europe, Data and Results. Springer Verlag Berlin,
    Heidelberg, and New York.

Helbig, K., (1981a), Anisotropy and dispersion in periodically
    layered media. Paper presented at the 51st Annual SEG
    Meeting, Los Angeles, CA.

Helbig, K., (1981b) Ray geometric migration in seismic prospecting.
    In: The Solution of the Inverse Problem in Geophysical
    Interpretation, edited by R. Cassinis, Plenum Press, New
    York and London: p. 141-177.

Hosken, J.W.J. (1981), Imaging the Earth's Subsurface with Seismic
    Reflections, In: The Solution of the Inverse Problem in
    Geophysical Interpretation, edited by R. Cassinis, Plenum
    Press, New York and London: p. 179-210.

Hubral, P., Krey, Th., (1980), Interval velocities from seismic
    reflection time measurements. An SEG Publication.

Hubral, P., (1980), Wavefront curvatures in three-dimensional
    laterally inhomogenous media with curved interfaces,
    Geophysics, Vol. 45, No.5: p. 905-913.

Kern, H. (1978), The effect of high temperature and high confining
    pressure on compressional wave velocities in quartz-bearing
    and quartz-free igneous and metamorphic rocks.
    Tectonophysics 44: p. 185-203.

Krey, Th., (1976), Computation of interval velocities from common
    reflection point moveout times for layers with arbitrary
    dips and curvatures in three dimensions when assuming small
    shot-geophone distances. Geophysical Prospecting 24:
    p. 91-111.

Krey, Th., (1978), Seismic stripping helps unravel deep re-
     flections. Geophysics, Vol. 43, No.5: p.899-911.

Krey, Th., (1980), Mapping non-reflecting velocity interfaces by
     normal moveout velocities of underlying horizons.
     Geophysical Prospecting 28: p. 359-371.

Krey, Th., (1980), Straightforward derivation of Hubral's
     wavefront curvature differential equations in inhomo-
     geneous media. Short Note, Geophysics, Vol. 45, No.5:
     p. 964-967.

Lynn, W., Claerbout, J. (1979), Velocity estimation in laterally
     varying media. Paper presented at the 49th Annual SEG
     International Meeting, New Orleans, Louisiana.

Meissner, R., Bartelsen, H., Krey, Th., Schmoll, J. (1980),
     Combined reflection and refraction measurements for
     investigating the geothermal anomaly of Urach.
     In: A.S. Strub and P. Engemach(Editors), Advances in
     European Geothermal Research. D. Reidel Publishing
     Company, Dordrecht, The Netherlands: p. 1086.

Meissner, R., Bartelsen, H., Krey, Th., Schmoll, J. (1982),
     Detecting velocity anomalies in the region of the Urach
     geothermal anomaly by means of new seismic field
     arrangements. In: Geothermics and Geothermal Energy.
     Editors: V. Cermak and R. Haenel, E.Schweizerbart'sche
     Verlagsbuchhandlung, Stuttgart: p. 285-292.

O'Doherty, R.F., Anstey, N.A., (1971), Reflections on Amplitudes.
     Geophysical Prospecting 19: p. 430-458.

Taner, M.T., Koehler, F., and Sheriff, R.E. (1979), Complex
     trace analysis. Geophysics 44: p. 1041-1063.

GEOELECTRICAL DEEP EXPLORATIONS

BY MEANS OF DIRECT CURRENTS

Luigi Alfano

Institute of Geophysics
University of Milan
Via L. Cicognara 7, 20129 Milano

SUMMARY

The magnetotelluric method gives some doubtful results when
lateral variations occur. In that case the use of d.c. deep geoelectric soundings may be useful, owing to the larger number of data
which may be gathered by means of artificial currents in areas with
complicated structures, at least within 15 km depth. The possibility
has been shown of using the polar dipole-dipole arrays formerly ex-
perimented by the author for the execution of deep sounding in the
regional explorations.

INTRODUCTION

It is known that the aim of the geoelectric prospecting methods
concern the distribution of the electric resistivities underground,
and that the maximum possible exploration depths depend on the particular method used. This aim, which may be reached by means of
mathematical analysis of the measured data, may be interpreted in
terms of geological structure if other types of data are available.

The methods differ from one another according to the earth's
currents utilized, which may be direct or time varying, natural or
artificial. The possible exploration depths range from a few metres
to some hundred kilometres.

The electrical resistivity values of the rocks depend, in the
upper part of the earth's crust, on its chemical composition, but
also on the quantity and on the characteristics of the water contained
in the pores. These values range from a fraction of one ohm.m, to

several thousands of the same unity. The temperature may also influence the said values, particularly when electrolytic solutions occur. In the lower part of the crust the resistivity values are generally very high. In the mantle the same values are a descending function of the depth, as a consequence of the increasing temperatures.

In the upper part of the crust (about in the first 15 km), the geoelectric methods may give good results in distinguishing the unconsolidated, especially clayish (very conducting), rocks, and the moderately conducting marls, from many competent rocks, such as the igneous and the limestones. But in general the connections between the resistivity values and the different lythological types are not simple, and they vary for the different explored areas.

The maximum depths (some hundred kilometres) may be reached by the method known as "Geomagnetic Sounding". It is based on the recording of the time varying components of the geomagnetic field, which may be separated in an external (primary) part caused outside of the earth, and in an internal (secondary) part caused by the earth's currents induced by the primary current in the conducting earth. Now the ratio R = (internal field)/(external field) depends on the electrical earth's conductivity, which may be determined if the said ratio is known. As far as the penetration depth is concerned, it may be remembered that, according to the Maxwell's equations, and particularly to the well known "Skin Effect", shorter periods penetrate to shallow depths, and longer periods penetrate deeper. It shows that the calculation of the said ratio R for different intervals of the period spectrum permit the calculation of the resistivity as a function of the depth.

The computation of the ratio R may be attempted by means of the results of many stations distributed over the explored area, applying th spherical harmonic analysis.

The results obtained by means of this method, are affected by a noticeable amount of uncertainty, but they are practically irreplaceable in studying the mantle resistivities, which according to "geomagnetic sounding" carried out world-wide (between about one hundred and one thousand kilometres), decrease from about some hundreds to some hundredth ohm.m.

A somewhat different approach, used successfully to estimate the electrical conductivity between depths of about 10 and 150 km, is the magneto-telluric method. The variation of the overall horizontal components of the magnetic field are recorded during a time interval at a station. At the same station, and in the same time interval, the components of the geoelectric field are also recorded by means of probes placed into the ground. The variations of the two fields are obviously connected by the Maxwell equations: particularly, a knowledge of the ratio Z = "geoelectric field/geomagnetic field" makes possible the calculation of the underground actual resistivity in homogeneous media, and of the "apparent resistivity" in inhomogeneous media. According to the already mentioned skin-effect, the knowledge of Z for different periods, which generally range between 20 and 1000 seconds, makes possible the calculation of the resistivity as a function of the depth in the case of horizontal stratification.

Obstacles in the magneto-telluric method arise when the under-ground resistivity values depend not only on the depth, but also on the two horizontal coordinates, namely when lateral structural vari-ations occur. In such cases the information may be insufficient and also distorted, because the direction of the primary field is unknown and out of our control; moreover the same direction may change during the execution of a sounding, which, for great depths requires a long time. These difficulties became more harmful for depths smaller than 10 km, where the possible geological irregularities are closer to the observation points, and where the lateral variations may cause larger distortions in the recorded data.

Another difficulty of this method is the necessity of recording very small field values; this problem arises particularly for the magnetic measurements. The most modern devices used for this last type of measurement are the "cryogenic magnetometers" characterized by very high sensitivity, but also by high costs.

Modern instrumentation uses digital recording of data, namely of the two components of the electric field, and of the three com-ponents of the magnetic field; the recorded data may be introduced in a computer for a rapid and complete mathematical processing.

Some easy considerations on the said characteristics of the magneto-telluric method lead to the conclusion that it is a worth-while attempt to experiment with geoelectric soundings within 10 - 15 km depth using artificial currents. In effect the main advantage should be the possibility of controlling the positions and directions of the geoelectric fields in a given explored area. Moreover the possibility of carrying out the measurements with different fields having different poles in arbitrary positions, means that a much larger number of data, with respect to the magnetic telluric method, may be available. In a situation characterized by lateral variations this fact implies a very important advantage.

As far as the disadvantages are concerned, we will examine these in the next chapter.

THE DIPOLE-DIPOLE ARRAY FOR DEEP EXPLORATIONS

The most popular geoelectric method, using artificial currents, based on the Schlumberger array may be used in practice only for maximum exploration depths of about 2000 meters, which may be obtained, in the most favorable case, using a total 6000m spread length. Reasons of practical character justify this statement particularly concerning the noticeable length of heavy cables to be used and the inconvenience caused by the current leakages. Consequently it is necessary to substitute the Schlumberger array for greater depths even if its apparent resistivity diagrams are the best to be obtained by means of artificial and in particular by direct currents.

On the other hand larger depths may be reached only by means of dipole-dipole arrays, since these require shorter cable lengths to be spread on the ground. But this type of array which appears to be very good if applied to theoretical cases with horizontal stratification of homogeneous layers, proves to be not very useful in actual geological situations. In effect all types of dipole-dipole arrays present too great a sensitivity to the lateral and particularly to the superficial variations; namely the apparent resistivity diagrams obtained in many actual geological situations may be strongly distorted and scattered, so that the interpretation in terms of depth exploration may be difficult and often impossible. If follows that a transformation of the field dipole-dipole apparent resistivity curves into the corresponding Schlumberger curves is necessary, since these last, as already said, present a smaller sensitivity to the said lateral variations of the actual geological structures.

The above transformation between dipole-dipole apparent resistivity diagrams of any type and the Schlumberger curves, is expressed, for both the directions, by well known formulas (Alpin 1950, 1958), (Patella, 1974, 1981). But unfortunately these formulae are valid only when plane horizontal stratifications occur, namely without lateral variations. It may be shown that in more general structural cases no transformation formulae exists (Alfano 1980), in the sense that the problem is characterized by a number of unknowns larger than the number of data. The only exception is constituted by the dipole-dipole array of polar and continuous type, for which the dipoles lay on a line, possibly straight, and are contiguous without any interval (Alfano 1974). It derives that only by means of this last type of array is it practically possible to carry on the soundings, and particularly the deep soundings, by means of artificial currents.

We already pointed out that the transition from a field dipole-dipole to a desired apparent resistivity Schlumberger diagram is an integration, and consequently an arbitrary additive constant is required. This fact should appear, at least from a pure mathematical point of view to be an insurmountable difficulty, but it has been shown (Alfano 1974) that in practice this fact does not constitute an obstacle, if a correspondence even if not perfectly univocal between dipole-dipole and Schlumberger diagrams is desired. The results of the transformation are satisfying when used in actual geological situations.

172

Now the formulation of the best way of getting the desired data, namely of the more convenient array to be used in the field is only the first step toward the possibility of carrying out deep explorations. In effect difficulties may be encountered in the execution of the voltage measurements, since for increasing depths and consequently distances between the two dipoles, the signal becomes smaller and the noises, caused by the telluric currents, larger, so that the signal may be completely masked and the usual voltage measurement techniques loose their effectiveness.

The way to be followed is based on the digital recording of the voltage between the electrodes of the potential dipole during a time sufficiently long. These data must be mathematically processed with the aim of separating the signal from the noise. This processing, which may also be a usual Fourier spectral analysis, is based on the fact that the noise is characterized by a continuous frequency spectrum, while the artificial signal has only one constant and rigorously determined period.

The choice of the signal period is based on the skin-effect. This means that the period must be sufficiently long to make negligible the induction phenomena. The current commutes each half period, but the field behaves practically as a direct one.

On the upper part of Figure 1 the polar dipole-dipole array of the continuous type is shown, while in the lower part the corresponding half-Schlumberger appear, whose meaning will be clarified later. In the dipole-dipole array M and N are the potential electrodes; A_i and B_i (where $i = 1, 2, \ldots\ldots$) are the current dipoles. Keeping the potential dipole MN fixed, apparent resistivity measurements are carried out successively, by means of the different current dipoles $A_1 B_1$, $A_2 B_2$, $A_3 B_3$ $\ldots\ldots$ $A_n B_n$.

A peculiarity of a continuous sounding is that the current dipoles are contiguous namely B_1 is coincident with A_2, B_2 and A_3 and so on. This procedure is needed in the actual geological situations. In effect a sounding carried out by means of only some current dipoles separated by unexplored intervals should give rise to apparent resistivity diagrams defined only by some sampling points interpolated for the reconstruction of the whole field curve. Now these points may be sufficient and they may have a meaning in terms of vertical stratification only in quasi theoretical cases, dealing with rigorously homogeneous layers. On the contrary, in the great number of cases, its values may be noticeably and irregularly influenced by lateral variations and particularly by near surface random geological irregular structures, so that the interpolation between the sampling points may cause wrong apparent resistivity diagrams. In that case the evidence of a possible regular layering may be very poor, and moreover the possibility of a transformation from dipole-dipole to half-Schlumberger data may be very difficult or

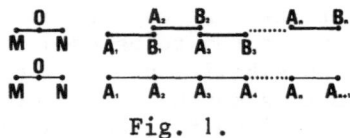

Fig. 1.

impossible. On the contrary the curves derived from the continuous array shown in Figure 1, even if influenced by the geological random irregularities, do not present losses of information, so that the transformation of the dipole-dipole into Schlumberger smoother curves is even possible.

The diagrams deriving from this transformation cannot be associated to a conventional symmetrical Schlumberger array, but to the half-Schlumberger one represented in the lower part of Figure 1; it has an asymmetric shape.

In Figure 2 we can see a theoretical example showing the transformation from a continuous polar dipole-dipole to a half-Schlumberger curve which may be needed if geological irregularities occur.

The upper and the lower part of Figure 2 concerns, respectively, Schlumberger and polar dipole-dipole diagrams; the continuous curves of both sections of the figure refer to the same structural situation, namely a horizontal stratification of homogeneous layers disturbed by some randomly distributed lateral variations. In the dashed curves we can see the effect of the vertical stratification only. It shows clearly that in the Schlumberger (upper) section of the figure the difference between the continuous and the dashed diagram is not so large to make impossible an interpretation in terms of vertical stratification, even if not very accurate. On the contrary, this impossibility is evident in the lower part of Figure 2, namely for the dipole-dipole curve.

In Figure 2 the values of the Schlumberger apparent resistivity, ρ_a and the polar dipole-dipole ones $\bar{\rho}_a$ are connected by the following relations:

$$\bar{\rho}_a(r) = \rho_a(r) \cdot \left[1 - \frac{1}{2} \frac{d \log \rho_a}{d \log r} \right] \qquad (2)$$

and

$$\rho_a(r) = 2r^2 \int_r^{\infty} \frac{\rho_a(r)}{r^3} \, dr$$

where r is the electrode spacing

In the practical calculations, since the field curves are represented by points, the formula must be substituted by the following
one:

$$\frac{\rho_a(A\mu)}{r_\mu^2} - \frac{\rho_a(A\nu)}{r_\nu^2} = \sum_{i=\nu}^{\mu} \frac{\rho_a(A_i, A_{i+1})}{r_i^3} \overline{\Delta r_i}$$

whose meaning is connected with the Figure 1. Particularly $\rho_a(A_\mu)$
and $\rho_a(A_\nu)$ are the Schlumberger apparent resistivity values when the
current electrodes are on the points and A_μ, and A_ν $_a(A_i, A_{i+1})$ is
the apparent dipole-dipole resistivity value obtained by means of
the current dipole $A_i \, A_{i+1} = \overline{\Delta r_i}$; and r_μ, r_ν are the distances OA_μ
and OA_ν; and last $\overline{r_i}$ is the distance $(r_i + r_{i+1})/2$ between the centers
of the current and of the potential dipoles. It derives from these
formulas that the transformation from the dipole-dipole to the
Schlumberger values is of integral type, so that an arbitrary constant
is needed; but it has been shown in the papers already cited that in
practice, this fact is not an insuperable difficulty.

The voltage measurements may be carried out in the following
way:

i) the energizing current is inverted regularly; the resulting period
has a minimum value of 60 seconds, and it is chosen in order to avoid
induction phenomena which cause the well known skin-effect. This
effect increases with the distance between the two dipoles and with
the apparent conductivity.

ii) if the ratio between the signal and the noise became too small
the analog recording of the voltages is not sufficient. In that
case the larger dynamical range of the digital recorder is needed:
the sampled values of the voltages (about twelve in a period), may
be printed on a paper strip, but, in more sophisticated instruments,
the magnetic recording and a computer make possible the analysis of
the data in the field in real time. By means of this analysis
carried out in the field it may be possible to determine the time
length of the record for each measurement without danger of mistakes.
In effect it is known that the length of the record must increase
for larger values of the noise/signal ratios.

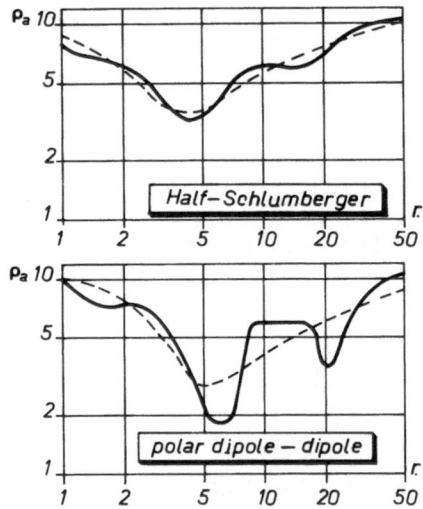

Fig. 2. Theoretical curves in the same disturbed geological
         situation.

Until now the technique of deep soundings carried out by means
of artificial currents has not reached its full potential and it may
be perfected. It is not easy to foresee now the maximum distances
between the dipoles which will be reached in the future.

In the author's experience, field cases may be remembered with
maximum distances of 25 and of 10 kilometers in situations with
apparent resistivity values respectively of 150 and 10 ohm.m. But
in these cases the energizing current was very low, namely no more
than 2 ampère. It was possible to get resistivity data from depths
ranging from 8 to 3 kilometers in areas of the Appennines (Italy)
namely in a region of very great interest and of complicated structure where the reflection seismic method is generally not effective,
and where the geology presents a high number of unknowns (Alfano,
1972; Patella 1979).

It will be possible to get better results by means of larger
energizing currents, longer record, and more efficient instrumentation.

The possibility of energizing the ground by means of high current
values depends clearly on the terrain conditions, since it is difficult to use powerful and consequently heavy generators if the
shape of the ground surface is too rough, and if roads or at least
some trails are lacking. Moreover the current values may be limited
by the contact resistance between the electrodes and the ground. In
effect low values of resistance often require time consuming oper-

ations, which must be repeated for each current dipole. A reasonable value for the current when distances between the dipoles exceed 10 km, and when low apparent resistivity occur may be about 10 ampère. Larger values may be used, when the very high conductivity of the surface formations permit sufficiently low contact resistances.

Another difficulty to be overcome is represented by the communications between the two crews working at the two dipoles. For the transmission of messages over long distances the popular walkie-talkie is not sufficient and often in too rough mountains it is inefficient.

Concluding, it seems that notwithstanding some field difficulties it is worthwhile to develop the use of this method, which, even for the first 10-15 km of the crust, may provide geoelectrical data hardly obtainable by means of natural currents, and particularly in those areas where the seismic methods fail.

REFERENCES

Alfano, L., 1974, A modified geoelectrical procedure using polar-dipole arrays. An example of application to deep exploration, Geophy.Prospec., vol.22, N.3:510-525.
Alfano, L., 1974, A first application of electrical dipole-dipole soundings to the calcareous formations in the central Appennines, Riv.Ital.di Geof., vol.23:101-117.
Alfano, L., Braga, G., and Marchetti, G., 1978, Considerazioni strutturali sulle finestre dell'Aveto e di Traschio alla luce di indagini geofisiche, Memorie Soc.Geol.Ital., vol.19: 461-468.
Alfano, L., 1980, Dipole-dipole deep geoelectric soundings over geological structures, Geophy.Prospec., vol.28:283-296.
Alpin, L. M., 1950, The theory of dipole sounding, in: "Dipole methods for measuring earth conductivity," Consultant Bureau, New York, 1-10.
Alpin, L. M., 1958, Transformation of sounding curves, in: "Dipole methods for measuring earth conductivity," Consultant Bureau, New York, 1866:61-77.
Patella, D., 1974, On the transformation of dipole to Schlumberger sounding curves, Geophy.Prospec., vol.22:315.
Patella, D., Rossi, A., and Tramacere, A., 1979, First results of the application of the dipole electrical sounding method in the geothermal area of Travale-Radicondoli (Tuscany), Geothermics, 8:111.
Patella, D., 1981, A general transformation system of dipole geoelectrical sounding into Schlumberger's, as an approach to the inversion, in: "The solution of the inverse problem in Geophysical Interpretation," Plenum Press, New York and London, 301,326.

REFLECTION PROFILING OF THE CONTINENTAL CRUST

Robert Phinney

Princeton University
Princeton, N.J., U.S.A.

EXTENDED SUMMARY

The COCORP program is devoted to collecting, analyzing, and interpreting high resolution seismic profiles using techniques which have been highly successful for hydrocarbon exploration in sedimentary basins. In six years of operation with a single conventional crew, over 3000 Km of section have been produced in I2 study areas. These profiles have sufficient resolution to show geologically informative structure at all depths in the crust. The high initial cost of data collection is more than justified by the geological value of the results.

The data area collected using the CDP field geometry. Typical field parameters are:

Source:	5 11 - ton vibrators sweep: 8-32 hz, for 30 sec, listen for 20 sec, 8 sweep per source point, arranged in an array
Sensors:	96 sensor channels (groups), spaced I00 m Source offset from channel 1: 600 m Maximum offset: I0.I Km geophone groups; 24 vertical geophones in a I00 m pattern, summed.
Line geometry:	Length: 50 - 500 Km per line Source point spacing: I00 m or 200 m Rate of progress: 2 - 6 Km per day

The data collection and processing procedures are based on having both source points and receiver points at a sufficiently small spacing that the wavefield can be sampled without aliasing. Consequently, signals are detected largely on the basis of their phase coherence: Initial processed sections follow the CDP (common depth point) protocol, which exploits the redundancy of coverage to enhance the weak reflection signals.
A typical processing protocol would be:

1. Demultiplex

2. Edit

3. Collect CDP gathers

4. Static corrections

5. Stacking velocity analysis

6. Normal moveout correction and stack

Filtering and deconvolution may be applied before and after stack; finally, to rectify a grossly distorted subsurface geometry, we add, despite its occasional drawbacks

7. Migration

The quality of COCORP sections varies quite a bit, from excellent to poor. In general, two problems arise which affect data quality:

a) Since the survey is on land, static errors occur (anomalies due to unresolvable near-surface geology), which need to be estimated and removed.

b) The source strength (5 vibrators) appears to be only marginally adequate for reflections from deeper than about 20 Km.

The features which are most easily seen are layered in some sense. Thus, we have seen:

a) Deep sedimentary basins (Precambrian) beneath the Phanerozoic cover

b) Layered crystalline complexes, either gneisses or gneisses with igneous intrusions

c) Major low angle thrust faults.
High angle faults are extremely difficult to image unless the data quality is excellent. Consequently any high angle faults which appear in the interpretation of deep reflection data are the inference of the interpreter.

For the areas surveyed to date, some of the more noteworthy results have been illustrated.

1. Hardeman Basin - Wichita Mts, Oklahoma.
A thick Proterozoic section (6 Km) was found below I.5 Km of Phanerozoic cover. That section is missing in the Wichita Mt uplift, indicating more than 8 km of uplift in the early Paleozoic.

2. Rio Grande Rift, New Mexico. Below 1 km of block faulted basin fill, a Precambrian granitic basement contains at 22 Km an extensive sheet-like magma body. A horizontal Moho complex about 1 sec thick is seen at about 30 km

3. Wind River uplift, Wyoming. The Wind River range is found to be uplifted at least I5 Km on a intracrystalline thrust which flattens out to 20° or less at 25 km depth

4. Southern Appalachians . The Paleozoic thrust belts are found to be underlain by a master decollement which deepens from 5 km in the west to over I0 km in the east (200 km distance). A shortening of the surface and a consequent translation of the overthrust plate is inferred, of magnitude greater than 200 km. In the subsurface, the line of the former continental margin appears as a series of major dipping complexes, running as deep as 25 km.
Thrust faults are ubiquitous.

5. Taconic thrust belt, New England. In addition to the shallowly rooted Cambro-Ordovician thrusts, we see that the Precambrian crystalline rocks of the Green Mtn anticilinorium are emplaced on a low angle thrust.

6. Michigan Basin. Beneath the 3 km deep broad saucer of Paleozoics, we find a narrow I0 km thick Proterozoic basin which correlates with the extension of the Keweenawan rift system inferred from gravity and magnetics.

7. Data from the Archaean of Minnesota , the Grenvillian of the Adirondacks, and central New Mexico show strongly layered, dipping complexes of events, which await migration prior to interpretation

8. Midcontinent geophysical anomaly, Kansas. A thick, structurally disturbed Proterozoic basin is seen beneath the thin Phanerozoic cover, and extending to depths greater than I5 km.

Reflection groups corresponding to the lower crust-Moho complex are ubiquitous in these surveys, if not of uniformly usable quality. These events, which appear from 7 to I2 seconds, are frequently near-horizontal, and in individual cases form a zone around

1 or 2 seconds thick (3 to 6 km). This lower crustal complex frequently underlies shallower zones with few reflections.
We thus infer that one or more of the following processes have been significant in producing this complex, near-horizontal layering.

1. High shear strains and associated recumbent folding during periods of tectonothermal activity

2. Intrusion of large gabbroic bodies

3. Burial of former oceanic crustal sequences

EVIDENCE FOR A STRONG CRUST-MANTLE BOUNDARY UNDULATION IN THE

TYRRHENIAN SEA FROM A REFRACTION SEISMIC SURVEY

Christel Böttcher

Institut für Geophysik, Universität Hamburg
Bundesstrasse 55
2000 Hamburg 13, BRD

INTRODUCTION

In 1980 the Institute of Geophysics, University of Hamburg and the Osservatorio Geofisico Sperimentale OGS, Triest were running some refraction seismic profiles in the Tyrrhenian Sea, the Ionian Sea, the Italian Peninsula and Sicily. Profile 5 was situated in the Tyrrhenian Sea off Naples and extended on land through to the Adrian Sea.

In this paper I like to talk about the profile in the Tyrrhenian Sea, the data involved and the problems which arose during the data processing.

Because we were interested in detailed informations about the crustal structure including the crust-mantle boundary, six seismic sensors with a spacing of ~20km were deployed. We used Ocean-Bottom-Seismographs (OBS) which were developed at our institute. These OBS consist of two main parts. A 4.5 Hz 3-component geophone and hydrophone external pack and the electronic and recording unit, connected to a surface buoy. The data are recorded directly on a twin 4-channel cassette tape recorder. The tape runs for at least 100 h, recording all events continuously. (For a more detailed description see Herber et al., 1981)

FIELD DATA

The 120 km long profile started in the abyssal plain with a water depth of 3400 m and ended 10 km off Ischia, water depth 1000 m, see Fig.1. Shot spacing was 1 km, corresponding to 1 shot every six minutes. To optain a high seismic energy range we used

183

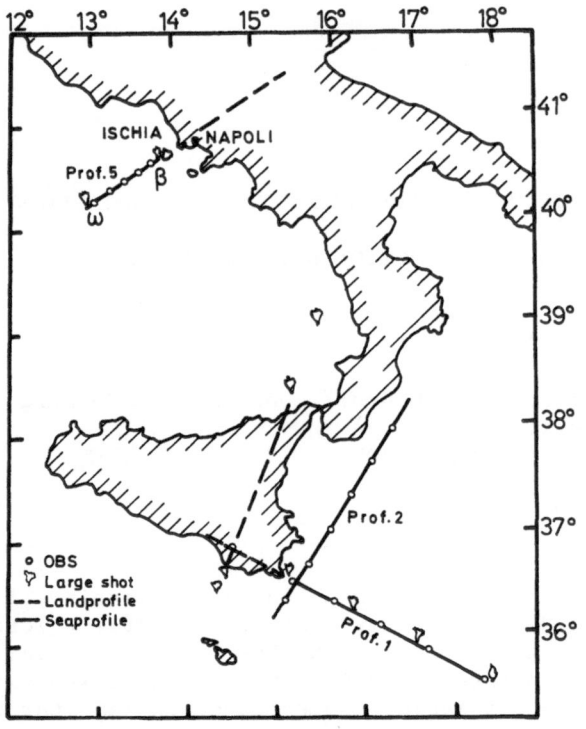

Fig.1 Location map

dispersed charges, fired every half and full hour: 2*25 kg
($\cong$ 1*100 kg) and 2*50 kg ($\cong$ 1*200 kg), normal charges were 1*25 kg.
Specially for the land stations we fired big charges on both ends
of the profile (3*150 kg $\cong$ 1*1350 kg).

We lost the OBS on position α, the one nearest to the coast,
but obtained good results from 85 shotpoints for the other five
OBS with data on 1 hydrophone and 3 geophone chanel.

Even with the aid of an integrated navigational system (Sate-
lite Navigation, Loran-C, Radar) it remains difficult to find the
true shot-point and receiver positions on seaprofiles. These difficulties result from the following problems:

1. While launching the OBS it will most probably not sink vertically.
2. While deploying the OBS the ship is exposed to drifting
   forces.
3. The distance ship-shotpoint is not always constant.

We reached an accuracy of ±200 m in calculating the true shotpoints and OBS positions from the ship navigation system and water
sound.

The shooting procedure was designed to generate a bubbleperiod
of 200ms, that means to generate a wavelet with the peak frequency
of 5 Hz. Herber and Snoek (personal communication) show, that a
seismic 5 Hz source would guarantee best energy transmission through
soft sediments, as a source below this value would lie in the range
of the increase of ambient noise, whereas for higher frequencies it
underlies dispersion.

SEISMOGRAM SECTIONS

In this paper I present seismogram sections based on analog
hydrophone data (see Fig.2). All seismogram sections are reduced
with 6 km/s. The restriction for shooting over a seacable crossing
our profile caused a 20 km wide seismic gap between OBS positions
ω and ε.

Characteristic for most seismogram sections are strong second
events, i.e. insets of seismic energy which was reflected at the
water surface before being recorded by the hydrophone and therefore
appeared parallel to the first arrivals in the seismogram sections.

The water sound, which is generated by chemical sources, is
characterized by large amplitudes with high dynamic ranges, ex-
ceeding the dynamic range of seismic energy by up to 60 dB. Such
high values overloaded the amplifiers of the recording units, making
it impossible to analyse any signals from tape except those signals
arriving at the receiver before the water sound. Up to a distance
of 7 km from the OBS, depending on the water depth, we recorded the
water sound as first arrival and therefore omitted the corresponding
seismograms.

MODELLING THE DATA

For solving the inverse problem I used a ray tracing program
written by Cerveny and Psencik. This program is designed for the
computation of rays, wavefronts and traveltimes of seismic body
waves propagating in lateral inhomogeneous media with curved interfaces.

There were some restrictions for the applicability of the program version we used, making it quite difficult to built a model
that could satisfy our measured data: e.g. blockstructures could
not be used; amplitudes were only calculated for the upper layer.

Our starting model was based on a first interpretational model gained from multichanel reflection seismics by R. Nicolich, Triest. From this profile we took the informations about the upper sedimentary layers which could not be resolved from our refraction seismic work. The OGS reflection and our refraction data yielded the following velocity distribution for p-waves: 2.8 km/s for the upper layer, 4.5 km/s and 6.0 km/s for layers 2 and 3, 8.0 km/s for the upper mantle. All traveltimes are reduced with 6 km/s.

A problem consisted in calculating rays for a given layer for precisely that range for which measured traveltimes existed, resulting from the complicated geometrical shape of the boundaries. You can solve this problem by setting the angle increment to the lowest possible value, causing the calculation time to increase rapidly. To avoid this we changed the program slightly thus allowing to assign several ray path sectors with different angle increments for each OBS (see Fig.3).

The iteratively changing of boundaries and velocities within the model just means trying to obtain the best fit of calculated and measured traveltimes (measured traveltime data are marked by symbols, calculated data are correlated through lines).

The first motion of the arriving seismic wave sometimes is very weak and therefore can vanish in the noise whereas second and consecutive oscillations generally arrive with higher amplitudes. Imagine a 3 Hz seismic wavelet arriving at the receiver: If you miss its first motion in the seismogram you will at least miss half a wavelength. In other words your estimated traveltime would be 150 ms too slow. For this reason a zero traveltime difference of calculated and measured data is accidentally. We stopped the iterations of boundaries, velocities and gradients after reaching an average accuracy of ± 150 ms of calculated and measured traveltimes.

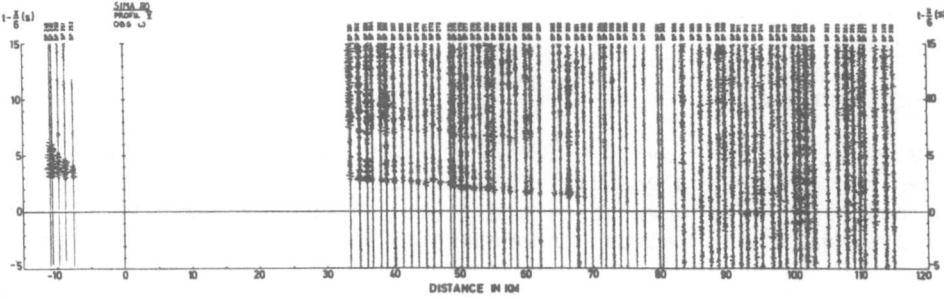

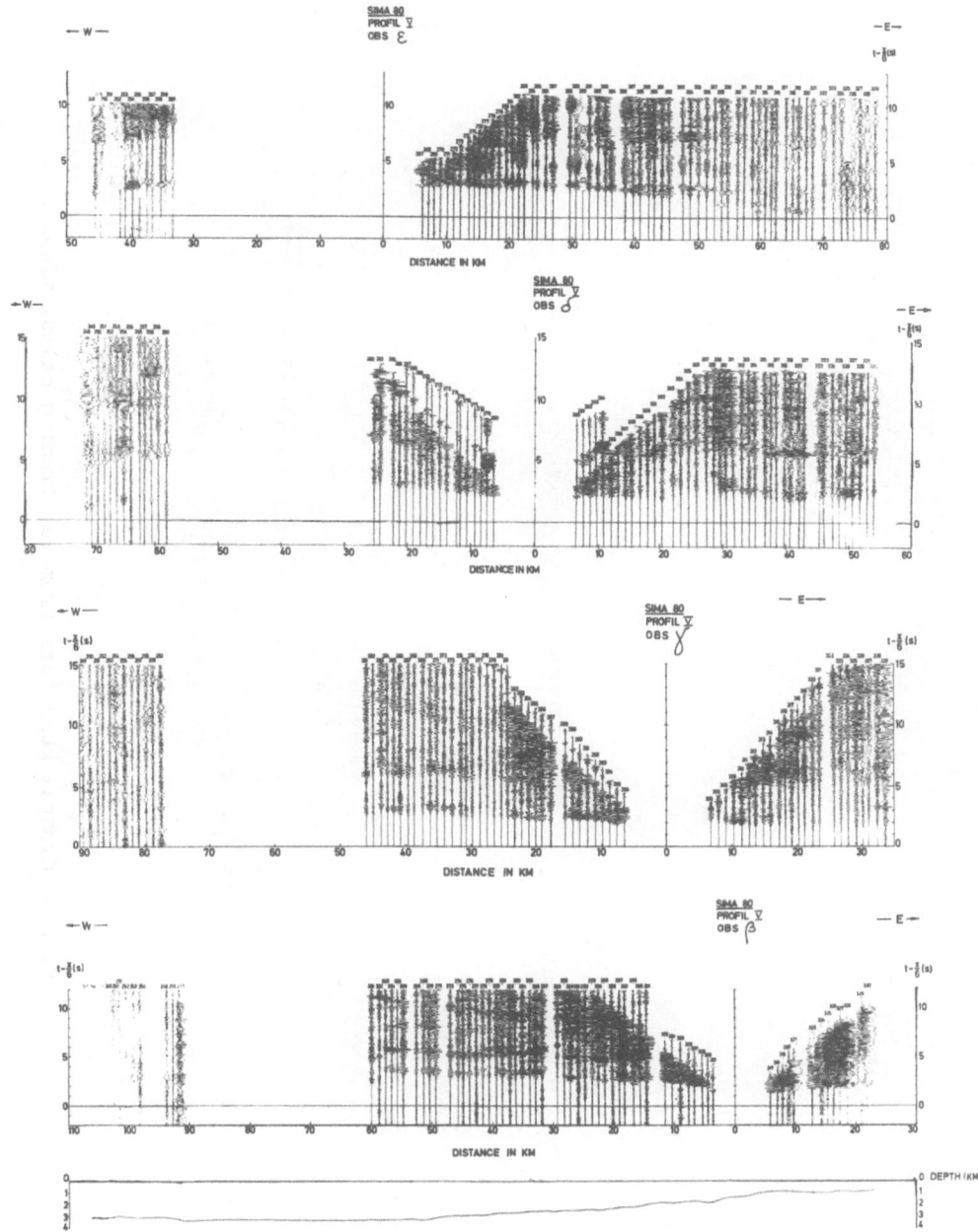

Fig.2   Seimogram sections

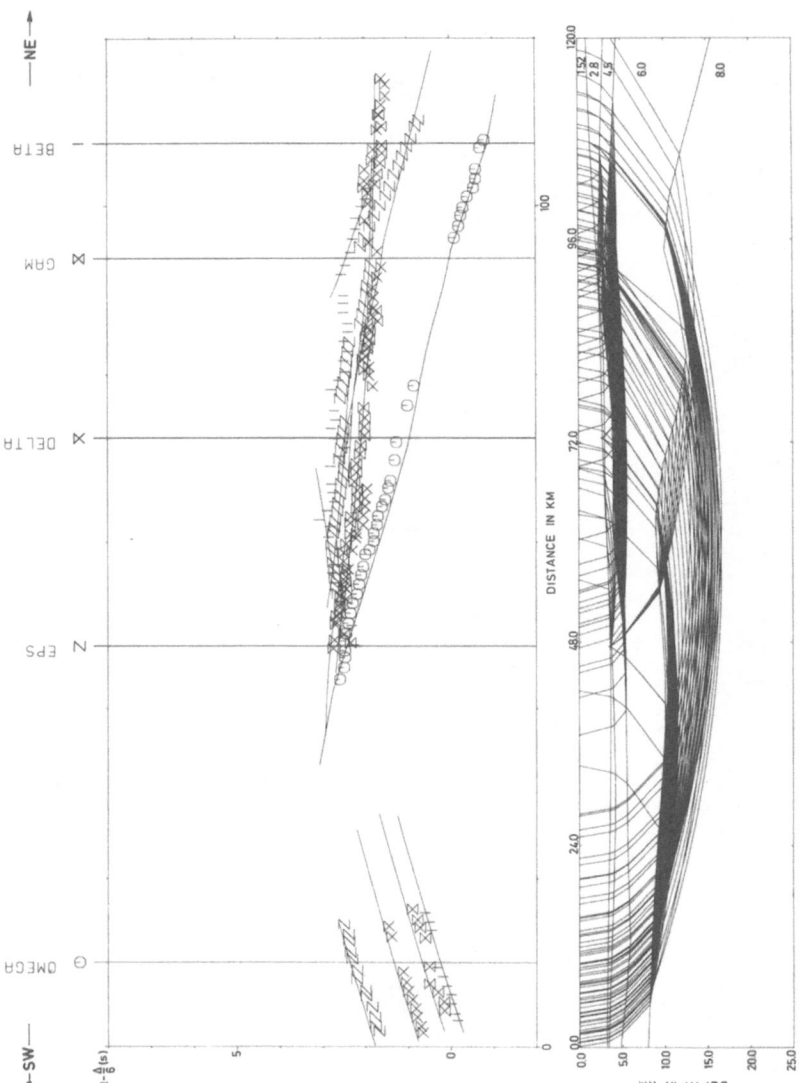

Fig. 3 Crustal structure from ray tracing calculations

188

RESULTS

As stated earlier, the experiment was designed to obtain a
detailed crustal model and to define the structure by multiple
coverage and reverse profiling.

A fairly precise location of the layer boundaries has been
achieved, as good data from 5 Ocean Bottom Hydrophones were
available.

The investigation of the mantle boundary showed two remarkable
features which sofar have not been seen in other seismic experiments:

1. Crustal thickness 200 km off Naples is about 6-7 km.
   This extremely low value was verified by "Moho" data obtained
   from OBS positions β, γ, δ and ε.

2. The shape of the "Moho" is characterized by strong undulations.
   In the abyssal plain the "Moho" rises up to 9 km under sea-
   level, dips downward to 14 km of depth in direction of Ischia,
   rises to a swell-like structure with 10 km of depth and then
   dives under the Italian Peninsula.
   A similar behaviour can be observed for the other layers on
   top of the "Moho".

Considering the pronounced position of the Tyrrhenian Sea in
the interaction zone of the African and Eurasian Plate it seems impossible to assign the crust to a certain type by only using the
seismic interpretation based on traveltime analysis and ray tracing
for hydrophone data.

We hope to obtain more information also for the petrographical
interpretation from the waveform analysis of the 3-component geophone data. However it should be pointed out, that for a final
geodynamical interpretation of this area results from landstations
must be incorporated.

REFERENCES

Herber, R., Nuppenau, V., and Snoek, M., 1981, An OBS system for
marine seismic investigations basic requirements and options:
The Hamburg OBS, Boll. di Geof. Teor. ed Appl.

Herber, R., and Snoek, M., 1982, Energy generation for refraction
seismic experiments at sea: Technology for chemical sources,
submitted to: Marine Geophys. Res.

A BENDING MODEL FOR THE CALABRIAN ARC

G. Gaudiosi[°], G. Luongo[°] and G.P. Ricciardi[°°]

(°)   Istituto di Geologia e Geofisica, Università di
      Napoli, Italy
(°°) Osservatorio Vesuviano, Ercolano, Napoli, Italy

INTRODUCTION

In recent years, many hypotheses have been proposed trying to explain the complex dynamics of the Calabrian arc and of the Tyrrhenian basin.

[1,2,3]According to the classical Plate Tectonics model, many authors thought this area to be an arc‑trench system which developed from the subduction of the Ionian plate in a NW-SE direction, beneath the Calabrian arc, taking as evidence the presence of intermediate and deep earthquakes in the southern Tyrrhenian sea.

This process caused the formation of a marginal basin: the Tyrrhenian basin, and a calc-alkaline volcanic arc: the Aeolian islands.

Recently, some researchers[4,5] have proposed an alternative model. In their opinion, the Tyrrhenian sea originated from a rifting of a deformated belt marking the Alps-Apennine collision suture zone, accompanied by an anticlockwise rotation of the Apennines and an eastward shifting of Sicily.

In this study, a different interpretation of the Calabrian arc dynamics will be discussed.

From our point of view, a plate in subduction should be considered as a passive element of the dynamics of the area, rather than as an active one.

According to our model, the present structure and dynamics of the arc is due to the strain of an elastic plate in shear-bending under the stress produced by the Africa-Europe collision, which started 10 My ago, when the Tyrrhenian sea began to open.

In our opinion, the shear-bending energy will be responsible for the faulting process that takes place in the arc and should be of the same degree of magnitude as the resulting seismic energy released.

In order to verify if this model accounts for the Calabrian arc dynamic processes, we shall briefly analize the tectonic characteristics, the earthquakes' distribution and the focal mechanisms of the area. Moreover, the seismic energy released in the arc and the shear-bending energy will be compared.

GEODYNAMIC PROCESSES

The Apennine chain forms in southern Italy a wide-arc convex in the direction of the Ionian sea.

The tectonic analysis of this area allows the identification of a series of longitudinal and transverse faults to the axis of the chain.

The longitudinal structures show a direction which changes from NW-SE in Campania and Basilicata, to N-S and NE-SW in Calabria and finally to E-W in Sicily. The azimuthal change of the direction occurs at the points where the longitudinal structures cross with the transverse ones.

The volcanic picture of this area, as already shown by Barberi[2] and Marinelli[6], is characterized by many active or recent magmatic activities with a geodynamic aspect which differs greatly from zone to zone. It changes from abyssal tholeites in the center of the Tyrrhenian sea, to K and high-K volcanic rocks in the Neapolitan area, to calc-alkaline magmatism in the Aeolian islands and finally to an intraplate-type distensive volcanism in the eastern Sicily (Etna, Iblei), and Pantelleria graben.

The earthquakes' distribution shows that the main shallow activity is concentrated along the chain.

The biggest part of the seismic activity is recorded along the NE-SW alignment which extends from Catanzaro trough to southern Sicily through the Messina Straits and the Etnean volcanic area.

The seismicity of the southern Tyrrhenian sea, as shown by many authors[1,7,8], is characterized by intermediate and deep focus earthquakes, defining an almost continuous structure dipping about 50° in the NW direction and reaching a depth of 500 km.

The fault plane solutions of earthquakes occurring in the south-

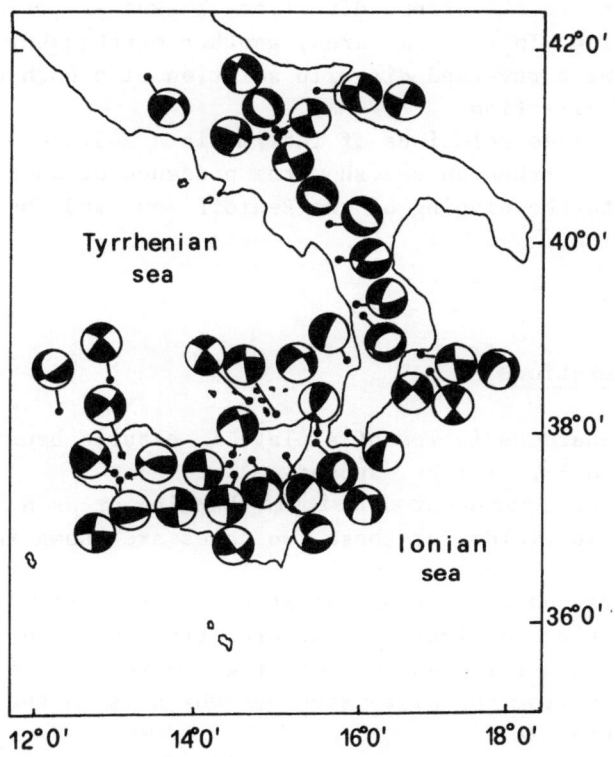

Fig. 1. Fault-plane solutions of some shallow earthquakes occurred in southern Italy from 1908. Data are projected on a Wolff net, lower hemisphere. White quadrants: dilatation; black quadrants: compression.

ern Apennines and the Calabrian-Sicilian region show a very complex picture.

The available mechanisms, determined by many authors[9,10,11,12], are shown in Fig. 1.

In the southern Apennines and in northern Calabria, strike-slip and normal fault mechanisms dominate with tensional axes which are parallel to the anti-Apennine direction.

In southern Calabria and in the Sicily region, the focal mechanisms show a very variable stress pattern.

In western Sicily, reversed dip-slip and strike-slip are observed. The T-axes are orientated in a N-S direction.

In eastern Sicily, the available data show a dip-slip motion along a NNE-SSW plane. Other solutions agree with transcurrent faults in the NE-SW direction.

In the Messina Straits, the 28-12-1908 earthquake indicates a normal faulting, in the graben direction. T-axes are orientated in a NW-SE direction. In the same area, another earthquake, further to the north, shows a reversed dip-slip solution with both planes running in a NE-SW direction.

The fault plane solutions of intermediate and deep earthquakes of the southern Tyrrhenian sea show the presence of down-dip compression parallel to the dipping of the Benioff zone and the T-axes in ESE-WNW direction.

BENDING MODEL

General considerations

The main characteristics of a plate in elastic bending, examinated by Johnson[13], will be exposed.

Such a process can occur as a simple bending or as a bending with shear. The strain fields for these two cases are shown in Fig. 2A and 2B.

In the first case, only normal stresses are applied to the plate and its deformation consists of fractures transverse to the length of the plate; a neutral plane divides the compressive zone which is in the innerside from the distensive one which is in the outerside of the plate (Fig. 2A).

In the second case, a shear stress is also present.

Therefore, as shown by Feodosev[14], longitudinal and transverse fractures are produced and the neutral plane position becomes the maximum shear zone and maybe the more fractured zone of the plate.

By fixing a bar at one of the two ends, this process is usually exemplified (Fig. 2B). In this case, both normal and shear stresses develop on the plate transverse section that don't remain flat, as usually happens in a simple bending process.

The strain field resulting from a shear bending process appears somewhat complex. In fact, in this case, besides simple fractures,

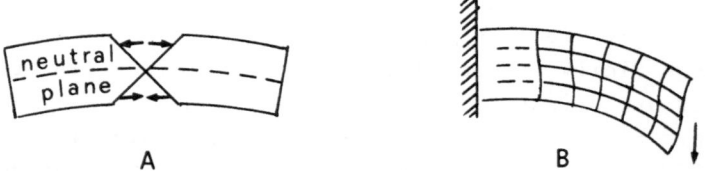

A   B

Fig. 2. Strain field in an elastic plate under: A) a bending simple stress; B) a bending stress with shear.

transcurrent movements may be produced between the transverse and longitudinal fractures. The unevenness of the surface, along which transcurrent motion takes place, may produce a series of local leaky transform faults both longitudinal and transverse to the strike of the plate. The complexity of the strain field increases at the part near the fixed end of the plate, where maximum shear accumulates.

The comparison between the bending models and the tectonics of the Calabrian arc shows that the shear bending process fits better the dynamics of the region.

Dimensions of the plate

From tectonic and seismological evidences we make the hypothesis that the plate submitted to the bending process extends approximately from 36°30' N to 42° N; the northern boundary runs from the Adriatic to the Tyrrhenian sea along a N-S large belt which covers a part of the central Apennines. This is a transitional zone between the central-northern and southern part of Italy; in fact, here, occurs an inversion of the bending of the Apennine chain and a change of the seismic characteristics.

We think that, from the beginning of the Tyrrhenian opening, the southern part of the peninsula bent more than the northen part and shifted eastwardly faster. Consequently, two arc-structures developed with partially independent evolutions.

Finally, we suppose that the southern free end of the arc runs along the Pantelleria graben.

The other dimensions of the plate have been fixed on the basis of the distribution of the shallow earthquakes occurring in the bent area, because it is believed that the Tyrrhenian intermediate and deep earthquakes are not caused by the bending process, but are related to a relic Benioff zone, which is partly detached from the upper part due to a decoupling of the slab.

Therefore, we consider the dimensions of the Calabrian arc to be the following:

T = 80 km   thickness
a = 60 km   width
L = 600 km   length
ρ = 200 km   radius of curvature

Analysis of bending energy

On the basis of the previous considerations, we shall calculate the shear energy of bending process and the seismic energy which is released in the Calabrian plate.

According to Feodosev[14], in a bending process, the normal stress is:

$$\sigma = \frac{M}{J} y = \frac{B}{\rho} y \qquad (1)$$

where M is the bending moment, J the inertial moment, y the semi-width of the plate, B the Joung's modulus for a plate. In a rectangular section of a plate, in the case of a bar fixed at one end, the shear stress is:

$$\tau = \frac{\sigma T}{4 L} \qquad (2)$$

where L and T are the length and the thickness of the plate, respectively.

The shear stress energy is:

$$W_{sh} = \frac{1}{2} \int_0^V \frac{\tau^2}{G} \, dV$$

where V is the volume of the plate and G is the modulus of rigidity.

Assuming the Joung's modulus $B = 4.3 \cdot 10^{11}$ dyne/cm^2, it follow that:

$$W_{sh} \cong 4 \times 10^{28} \text{ erg}$$

Analysis of seismic energy

The bending process, probably, started 10 My ago, when the Tyrrhenian sea began to open. Assuming that the seismic energy has been released in the past with the same rate as it is released at present, since the stress field is considered to be constant in time, therefore, the strain released for the last 10 My is calculated.

Since the total seismic energy released in the Calabrian arc from 1900 to 1975 corresponds to:

$$W_s = 6.5 \times 10^{22} \text{ erg}$$

therefore, the energy released in 10 My is:

$$W_s \cong 10^{28} \text{ erg}$$

This value fits well the shear bending magnitude.

CONCLUSIONS

The stress-strain field for the southern Italy, inferred from
the geological and geophysical available data, agrees with those of
a bending plate fixed at one end as it is outlined in Fig. 3. In fact:
- The acting stress field in southern Italy is prevalently compressive in the innerside of the arc as supported by the occurrence of
  the Aeolian andesitic volcanism and the focal mechanisms solutions.
- The focal mechanisms, the alkali-basaltic volcanism of Etna and
  the presence of graben structures in the Calabrian and Sicilian
  area testify a distensive stress field in the outerside of the arc.
- A series of fractures parallel and transverse to the axis of the
  arc are present, along which seismicity is mostly concentrated.
  Moreover, there is no evidence of an aseismic neutral plane as pos-
  tulated by the model of a simple bending, but the seismicity is at
  its highest level along the axis of the plate, where the shear

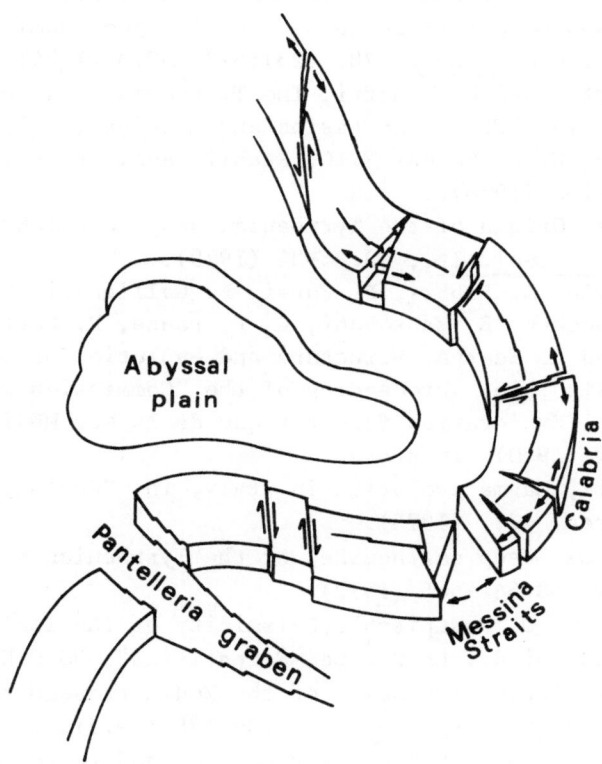

Fig. 3. A simplified dynamic model of the Calabrian arc.

stress and the fracturing process are maximum, in agreement with the model of a bending with shear.

- In the northern part of the arc, closer to the bound end of the plate, normal faults are predominant, determined by leaky transcurrent movements between the longitudinal and transverse structures. This strain field is a result of the focal mechanism solutions and the Campanian volcanic structures. The latter are also indicators of the regional stress field as shown by Gaudiosi[15].
- The conclusion, which can be drawn with the above study, is that there is a perfect agreement between the energy of the shear bending process and the seismic energy released in the arc.

REFERENCES

1. M. Caputo, G. F. Panza, and D. Postpischl, New evidences about deep structure of the Lipari arc, Tectonoph., 15:219-231 (1972).
2. F. Barberi, P. Gasparini, F. Innocenti, L. Villari, Volcanism of the southern Tyrrhenian sea and its geodynamic implications, J. Geoph. Res., 78, (23):5221-5223 (1974).
3. M. Boccaletti, and P. Manetti, The Tyrrhenian sea and adjoing regions, in: "The ocean basins and margins", vol. 4B, A. E. M. Nairn, W. H. Kanes, F. G. Stehli, eds., Plenum Publishing Corporation (1978).
4. P. Scandone, Origin of the Tyrrhenian sea and Calabrian arc, Boll. Soc. Geol. It., 98:24-34 (1979).
5. G. Calcagnile, A. Fabbri, F. Farsi, P. Gallignani, C. Gasparini, G. Iannaccone, E. Mantovani, G. F. Panza, R. Sartori, P. Scandone, and R. Scarpa, Structure and evolution of the Tyrrhenian basin", in: Proceeding of the "Commission Internatiole pour l'Exploration Scientifique de la Mer Méditerranéé", Cagliari (1980), in press.
6. G. Marinelli, Magma evolution in Italy, in: "Geology of Italy", H. Squyres, ed., (1975).
7. A. R. Ritsema, Deep earthquakes of the Tyrrhenian sea, Geol. in Mijnbouw, 58:541-545 (1972).
8. M. Caputo, and D. Postpischl, Seismicity of the Italian Region, in: "Quaderni de: La Ricerca Scientifica", 90 C.N.R., (1975).
9. D. McKenzie, Active tectonics of the Mediterranean region, Geophys. J. R. Astr. Soc., 30:109-185 (1972).
10. M. Riuscetti, and R. Schick, Earthquakes and tectonics in Southern Italy, Boll. Geof. Teor. e Appl., 17:59-78 (1975).

11. F. D'Ingeo, F. Calcagnile, and G. F. Panza, On the fault plane solutions in the central-eastern Mediterranean region, <u>Boll. Geof. Teor. e Appl.</u>, 85:13-22 (1980).

12. C. Gasparini, G. Iannaccone, P. Scandone, and R. Scarpa, Seismo tectonics of the Calabrian arc, <u>Tectonoph.</u>, (1982), in press.

13. A. M. Johnson, "Physical processes in Geology", Freeman, Cooper & Company, San Francisco, (1970).

14. V. I. Feodosev, "Strenght of materials", Mir. ed., (1977).

15. G. Gaudiosi, G. Luongo, G. P. Ricciardi, Strain field in the volcanic areas of southern Italy and tectonics of the region, IAVCEI Symp., Tokyo, (1981).

A METHOD OF HYPOCENTRAL LOCATION BY MEANS OF AN EARTHQUAKE'S

FIRST MOTION

Marco Gasperini and Bruno Alessandrini

Istituto Nazionale di Geofisica
Via Ruggero Bonghi, 11/B
00100 Rome, Italy

ABSTRACT

By exploiting the possibility of determing the angle of emergence and the azimuth of a seismic ray at an observation point equipped with one vertical seismograph and two horizontal ones and using recent digital data acquisition techniques, an effective method for locating the focus of a seismic event is proposed. By means of this method it is possible to obtain, in addition to the point customarily considered as the "barycenter" of the focal volume, also a quantity which represents the "convergence" of various seismic rays on the source zone.

This method makes it possible actually to "determine" the focal parameters and does away with the need for approximation methods of the least squares type. In the present investigation, tests have been carried out to determine the accuracy of the angle of emergence and the azimuth for a number of earthquakes, together with other theoretical tests for the relocation of the focal parameters using a three-dimensional crust model.

INTRODUCTION

The problem of the space-time location of an earthquake's focus generally takes the form of a search for a point having coordinates x,y,z and t to which a physical significance may be assigned. This approach becomes a serious obstacle when it comes to attempting to re-establish the connection between the geological aspects and the physico-mathematical ones required for a description of the source of a seismic event.

It has been found, in practice, that the space coordinates of the focus alone may be determined from the data obtained from a single observation site provided that the value of the ground displacement of the longitudinal wave impinging on the earth's surface is known accurately enough. In other words, the records of the three components of a seismic station must show the beginning of the longitudunal and transverse wave sharply.

This approach, which has in the past been successfully used by all seismic observatories to locate an earthquake's focus, i.e. the starting point of the propagation of the perturbation caused by the sudden breakdown in the equilibrium under the earth's crust, now allows us to introduce a method for defining the source also in cases that cannot be treated as a mere geometric point.

This problem was in fact solved in the search for a point that, at a pre-established time, generates an initially spherical, isotropic, progressive condensation wave; the propagation of the disturbance takes place in an elastic, homogeneous, isotropic and suitably stratified medium.

Let us suppose that the wave front incident on the surface is recorded by a network of seismological stations equipped with the three ground displacement components at each recording site.

The value of the two angles, together with the arrival times of the first impulse at each station in the network, make it possible to describe the previous behavior of the longitudinal wave to compute, in a new way, the hypocenter of an earthquake.

DEFINITION OF THE FOCAL PARAMETERS

The earthquake focus is normally determined from the values of the arrival times of direct and refracted longitudinal waves. The only necessary condition for a solution to be found is that at least four first arrival observations be known in four observation points. In practice, however, such a condition is mostly found to be insufficient. The time read off the seismograms are affected by errors and the travel time diagrams used may not reflect the true velocities of the longitudinal waves in the case of the profiles under examination.

If the number of first arrival readings is greater than four, computing programs based on least squares type approximation techniques are generally used. In this case it is possible to assign to the solution a statistical error to allow its reliability to be evaluated with regard to experimental data scatter.

The solution is obtained by assigning space-time coordinate values to a starting point using specified criteria. Then, by means of successive corrections of these values, the deviation between theoretical and experimental arrival times at the various seismic stations is reduced to a minimum.

Unfortunately, there are serious drawbacks to this method. Generally speaking, starting from different points leads to different results being obtained. The latter are, in any case, heavily dependent on the type of crust model used.

As we have seen, an observation point equipped with a three-component seismometer can be used to obtain, in addition to the arrival time of the first impulse, also the values of two angles from which the ground displacement of the condensation wave can be determined. Of course, there is an increase in the overall amount of information. This allows us to make substantial modifications to the method used to locate the focus of a seismic event. It is, of course, possible to make an estimate, i.e. to start from a trial point and gradually correct to reduce the residues, which in this case consist of the differences between the theoretical and experimental values of the times and angles. In this paper a different method has been proposed, by means of which it is possible to define not only a point but also a quantity linked to the convergence of the various seismic ray on the source zone.

The method consists of tracing back the path followed by each ray using a three-dimensional crust model (taking into consideration also the lateral heterogeneities). The model is, in fact, made up of various sided suitably stratified "blocks" of various sizes, each having a different velocity. In passing through different "blocks", the seismic rays are subject to the laws of optical geometry. In order to take into account the error in each angle, four rays and not just one, are generally used for each seismological station. They form the vertices of a pyramid, the directions of which are defined by $A + dA$, $e + de$; $A + dA$, $e - de$; $A - da$, $e + de$; $A - dA$, $e - de$, where A and e are respectively the azimuth and the emergence angle and $2dA$ and $2de$ the standard deviations of the two angles.

The first step was to follow the various seismic ray of each observation site until the time in which the station nearest the source recorded the first shock. From each dt there is, generally, a different spatial distribution of the points representing the wave front at different points in time.

The "volume" obtained from each distribution will tend to decrease on approaching the source, and in general to increase and diverge as it moves away from it. Steps were then taken to determine the configuration corresponding to the "minimum volume" at a certain point in time representing the origin time of the seimic event and its "barycenter".

Table 1. The true values, the ones determined by our method and those determined by the first arival method, for each x,y,z and t, are reported in three columns. Obviously, the true values for the origin time t are set up to zero without reporting the corresponding column. In the last column, then, there are the values of the convergence.

x			y			z			t		c
7.50	7.51	8.70	17.50	17.62	18.40	6.00	5.49	1.00	-.05	.40	9.982
7.50	7.54	4.70	15.50	15.37	15.60	2.50	4.56	7.20	-.23	-.90	3.947
7.50	7.41	4.80	13.50	13.48	13.80	1.00	3.69	3.70	-.34	-.60	6.592
7.50	7.53	6.10	11.50	12.15	11.20	0.50	3.22	0.10	-.32	.50	7.941
7.50	7.50	8.10	9.50	9.50	11.10	5.00	5.81	9.10	-.15	-.40	.446
10.50	10.41	10.90	17.50	16.91	16.90	2.50	4.40	2.10	-.09	.10	9.994
10.50	10.74	10.90	15.50	15.19	14.70	4.00	4.75	1.80	-.19	.20	7.757
10.50	10.33	10.30	13.50	13.36	12.80	3.50	3.74	7.10	-.21	-.50	9.354
10.50	10.81	11.60	11.50	11.79	12.30	1.00	2.71	2.40	-.28	-.10	8.177
10.50	10.39	10.00	9.50	9.90	11.60	2.50	3.17	0.90	-.24	.40	4.968
13.50	13.79	11.30	17.50	16.72	14.60	4.00	4.08	8.30	-.04	-.50	7.733
13.50	13.83	11.00	15.50	15.45	12.40	5.50	6.98	2.10	-.23	.50	11.348
13.50	13.53	12.30	13.50	13.12	15.30	3.00	3.73	5.80	-.22	-.30	9.497
13.50	13.76	15.30	11.50	11.46	10.70	2.00	3.79	7.70	-.23	-.90	6.393
13.50	13.09	14.50	9.50	9.51	9.20	3.50	4.51	0.10	-.30	-.10	3.524
16.50	16.33	14.20	17.50	17.53	16.80	3.00	2.62	2.70	-.07	-.50	1.696
16.50	16.65	16.80	15.50	15.24	14.10	1.50	1.56	3.40	-.10	-.40	2.748
16.50	16.35	14.30	13.50	13.41	13.80	4.00	3.39	5.90	-.01	-.40	2.287
16.50	16.35	18.10	11.50	11.86	11.40	5.00	5.54	7.00	-.17	-.40	3.741
16.50	16.87	14.50	9.50	9.91	13.90	0.50	3.99	1.80	-.30	-.40	11.296
19.50	19.62	20.60	17.50	17.75	17.10	2.00	3.11	3.70	-.27	-.40	5.902
19.50	19.13	19.70	15.50	15.33	13.80	2.50	2.99	4.60	-.11	-.50	2.268
19.50	19.16	20.40	13.50	13.58	13.70	0.50	4.40	4.70	-.23	-.70	13.664
19.50	19.48	20.50	11.50	11.40	12.60	2.00	2.22	2.40	-.17	-.50	2.309
19.50	19.90	18.80	9.50	9.47	10.30	4.50	5.16	3.00	-.29	-.50	2.505

In addition to the value of this "volume", which obviously depends on the errors associated with the times and angles, as well as on the model used, it is also possible to find a quantity which represents the "convergence" of the various rays on the source zone.

It is obtained by calculating the sum of the minimum distances between the "barycenter" of the focal zone and each seismic ray.

RESULTS OBTAINED USING A SIMULATION

In order to test the proposed method, a simulation was carried out to obtain values for the focal parameters previously introduced for a certain number of seismic events. A "block" model with different velocities was defined over an area of 10 km x 15 km. Eight observation points were distributed over this area and the theoretical arrival times and angles at the stations determined for 25 points in a given three-dimensional distribution of foci. In the simulation the values of the two angles at source were corrected successively until seismic rays with an emergence differing by no more than 500 meters from each observation site were obtained. The focal parameters defined in the previous section were then determined for the 25 events.

The results obtained are shown in Table 1.

CONCLUSIONS

An initial consideration may be made concerning the stability of the solutions obtained as the model varies.

This is very important for a local network in which the model plays an essential role. Furthermore, the use of this type of method to locate seismic events could make it possible to use the ratio between the quantity linked to ray convergence and the "volume" to evaluate the physical size of an earthquake source.

REFERENCES

Galitzin, B., 1910, "Sur la detérmination de l'epicéntre d'un tremblement de terre d'après les données d'une seule station seismique," CNRS, Paris.

Caloi, P., 1932, "Nuovo metodo per calcolare le profondità ipocentrali," La Ricerca Scientifica, Milano.

Cagniard, L., 1939, "Réflexion et réfraction des ondes séismiques progressives," Gauthier-Villars, Paris.

Proceedings of seminar on hypocenter location methods, 1978, Progetto
   Finalizzato "Geodinamica," November, 9-10, Milano.
Alessandrini, B., and Levato, L., 1980, "Progressive estimation of
   hypocenters and three-dimensional crust model," XXXIII, Annali
   di Geofisica, Roma.

SEISMIC WAVE ATTENUATION IN SOUTHERN ITALY CRUST:

SCATTERING FROM RANDOM HETEROGENEITIES MECHANISM

OR INTRINSIC Q VARIATIONS WITH THE DEPTH?

Antonio Rovelli

Istituto Nazionale di Geofisica
Osservatorio Geofisico Centrale
Monte Porzio Catone, Rome, Italy

ABSTRACT

Determination of the frequency dependence in the band 0.1–25 Hz
for apparent Q of the seismic waves in a range of a maximum of 150 km
from the epicentre of the Irpinia earthquake (November 23, 1980) has
been made using displacement spectrum ratios computed by strong-motion
accelerograms recorded in the region. The application of this method
excludes a different $Q=Q(f)$ behavior on the Tyrrhenian side when compared with the Adriatic side of the peninsula where a different
structure of the crust is known to exist. The resulting relation is
practically the same

$$Q(f) = 40 \ f \qquad (\ f \ in \ Hz \ )$$

for the stations to the left and to the right of the Apennines.
Seismic wave attenuation seems to depend on this kind of analysis
more from the heterogeneity degree of the crust and from the scattering mechanism than from structural variations with the depth.

Following the Aki hypothesis that there must exist a peak of
Q^{-1} versus frequency in tectonically active zones, a minimum in the
low-Q frequency band has been sought in the range 0.05–2.5 Hz. A
clear result in this sense has not been possible: only in six $Q(f)$
profiles does an evident minimum exist, while in nine cases the $Q(f)$
curves are monotonically increasing from the lowest observable frequencies; a further nine cases appear of uncertain interpretation.

INTRODUCTION

The frequency dependence degree of seismic wave apparent Q shows interesting correlations with the tectonical activity of the regions where the seismic waves have been recorded.

Reviewing a series of previous works (Aki and Chouet, 1975; Aki, 1980a; Hermann, 1980), Aki exhaustively demonstrates that the dependence of apparent Q on frequency is a general fact and is stronger the higher the tectonic current activity in the zone in which Q is measured. Aki (1980a and b, 1981) points out that in Japan differences have been found between various regions (in a zone in the Kanto district $Q \sim f^{0.8}$, while in other zones of the same district $Q \sim f^{0.6}$). Also the results of Fedotov and Boldyrev (1969) for the Kurili Islands and of Rautian and Khalturin (1978) for Garm (Central Asia) reveal an apparent Q dependence on frequency of the type $Q \sim f^{n}$ (with n equal to 0.6 and 0.5, respectively). Aki considers to be non-random the fact that at higher frequencies ($\sim$25 Hz), apparent Q is in all cases considered high and practically independent from tectonical activity, actually proving equal to Q measured by means of the surface waves, whereas at about 1 Hz the value of Q differs widely according to the region and shows an obvious correlation with tectonical stability.

The recent earthquake in Irpina (23rd November 1980) and the large number of strong-motion accelerometric stations of the ENEL network situated in the epicentral region made it possible to study apparent Q as a function of seismic wave frequency and to investigate eventual azimuthal Q variations in connection with structural variations in Southern Italy.

THE METHOD

Q estimates have been performed starting from the assumption that the spectrum of a seismic wave transmitted through a dissipative medium can be expressed as a function of the hypocentral distance R and the frequency f by means of the following formula

$$E(f,R) = S(f) \ F(R) \ \exp\left(- \frac{2 \pi f R}{v Q}\right) \tag{1}$$

where S(f) represents the source spectrum, F(R) is the geometrical spreading term, and v is the wave velocity in the medium. Beside smoothing the experimental displacement spectra of the triggered stations, and assuming for the sake of simplicity that the geometrical spreading may be written as $F(R) \sim R^{-v}$, the spectral ratio method (Console and Rovelli, 1981) applied to different stations pairs at distance R_{n} and R_{m} furnishes

$$Q(f) = \frac{2\,\eta\,\log e\,f\,(R_m - R_n)}{v\left[\log\dfrac{E(f,R_n)}{E(f,R_m)} - v\,\log\dfrac{R_m}{R_n}\right]} \tag{2}$$

By varying the stations two at a time, numerous patterns have been computed for $Q(f)$ throughout the area under analysis, in correspondence of different values of v, representing different kinds of seismic wave propagation: $v = 2$ is typical of body waves, and corresponds to spherical spreading, $v = 1$ characterizes surface waves propagation, and $v = 0$ represents the case in which geometrical spreading is neglected, and the attenuation of seismic wave energy is due entirely to the exponential term in (1). Specimens of $Q = Q(f)$ profiles in different areas rounding the epicentral region are shown in Figure 1: a practically linear growth with frequency can be observed.

RESULTS AND DISCUSSIONS

The $Q(f)$ profiles obtained from (2) for pairs of stations in the Tyrrhenian side were compared with the analogous results of the Adriatic side. $v = 1$ was selected as being the one showing the greatest stability throughout the frequency range considered. By averaging separately for the Adriatic area and for the Tyrrhenian side, a best fit using least squares technique gave

$$Q(f) = 40\,f$$

both east and west of the epicenter. This fact deserves special considerations. In the Dainty (1981) model

$$Q(f)^{-1} = Q_o^{-1} + \frac{2\,\pi\,f}{v\,g_o} \tag{3}$$

Q increases linearly with frequency when the intrinsic attenuation Q_o^{-1} is negligible with respect to the last term in (3). g_o is the turbidity coefficient: Dainty postulates it is practically constant with frequency in the body wave frequency band, depending solely on the number of scattering centers per unit volume and the cross section of the scatters themselves, which is practically determined by their physical size. The Irpinia result would seem to indicate that the scattering mechanism of the seismic waves have greater effect on apparent Q than the dissipative effects; and this means that this parameter is more dependent on the heterogeneity of the southern Italian crust than on the physical characteristics of the different layers comprising the upper lithosphere in the eastern and western parts of the area considered. A basically similar effect from the physical standpoint, although occurring in quite different circumstances, was found by Spencer et al., (1982). They observed that,

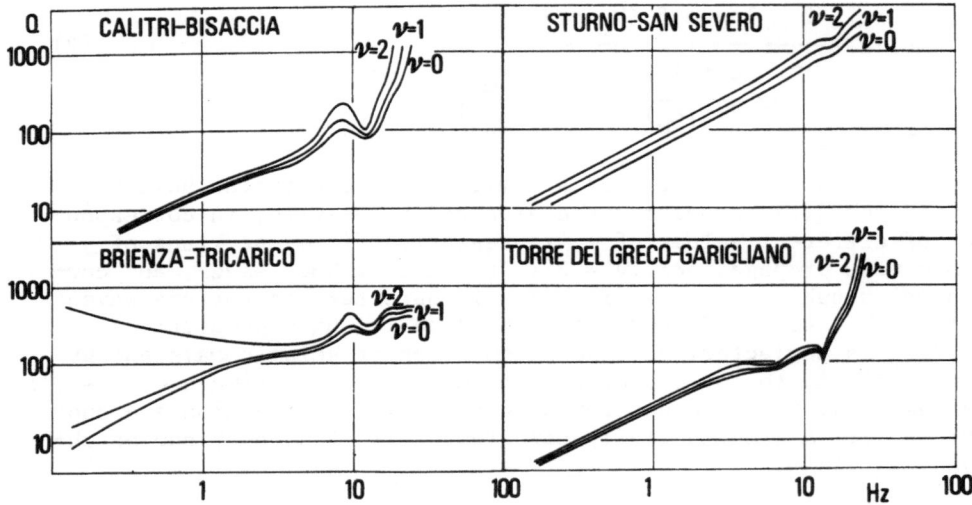

Fig. 1. Specimens of Q computed in the hypothesis of absence of
geometrical spreading ($\nu = 0$), and for surface ($\nu = 1$)
or spherical ($\nu = 2$) propagation of the seismic waves.

on analysing seismic data from seismometers situated at different
depths along vertical profiles, for small distances between seismometers, Q values computed by the spectral method were influenced more
by the effects of local stratigraphy than by the actual attenuation
undergone by seismic waves over the distance between the seismometers.

DETAILED ANALYSIS IN LOW-Q ZONE

One significant aspect of the dependence of Q from the tectonical
stability is its behavior at low frequencies: in tectonically more
active zones, at about 1 Hz, Q takes on much lower values than in
more stable zones: when it happens, Q must show a minimum between
about 0.2 and 1 Hz, since the apparent Q obtained from surface wave
analysis (e.g. at 0.05 Hz) is extremely high, i.e. between about 500
and 1000. Experimental profiles of apparent Q should, according to
Aki (1980a and b, 1981), thus display a minimum point for decreasing
frequencies, tending to increase again at the lower frequencies that
can be analysed using this method for values corresponding to the
characteristic band of the surface waves. In the preceding processing, the drastic smoothing of the spectra failed to estimate the Q
behavior in low frequency bands: high resolution spectra were then
used for a detailed analysis in the 0.05-2.5 Hz band. These Q(f)
curves appear of different types: not all covered a coherent profile.
Only six had an obvious minimum between 0.2 and 1 Hz for both $\nu = 1$
and $\nu = 2$. In many cases (nine in fact) the Q(f) curves are monotone

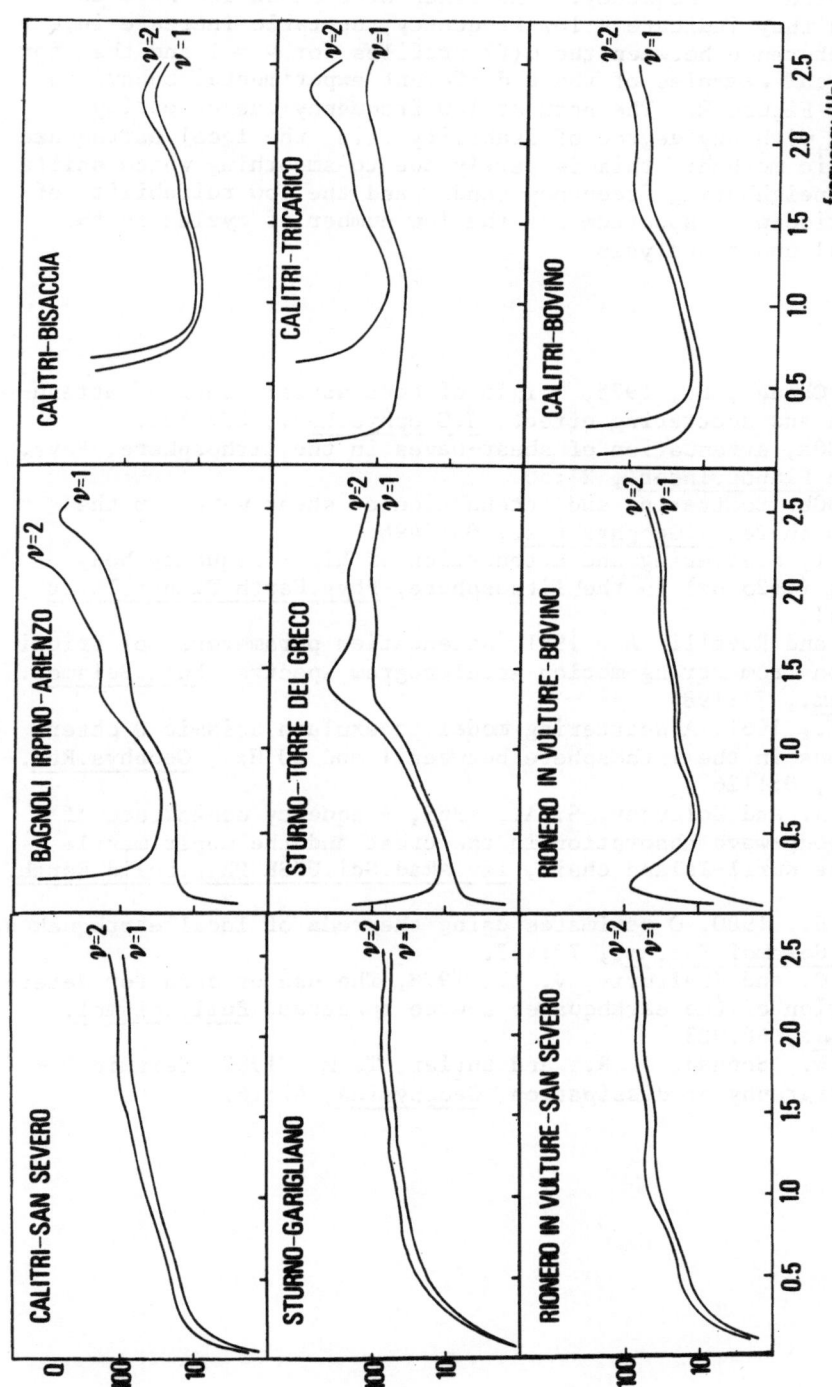

Fig. 2. "Zoom" in low–Q zone: on the left, the Q trend is monotonically increasing as a function of the frequency; in the middle, Q shows a minimum between 0.2 and 1 Hz; at right, some cases of difficult physical interpretation.

increasing with the frequency. The other nine cases are hard to interpret as they indicate a low-frequency, unstable increase in Q or else incoherence between the $Q(f)$ profiles for $\nu = 1$ and that for $\nu = 2$. Typical examples of these different experimental behaviors are shown in Figure 2. The peak at low frequency cannot easily be displayed with any degree of stability using the local earthquake spectral ratio method: this is partly due to smoothing which shifts energy into neighboring frequency bands, and the low reliability of the long period wave spectrum for the low number of cycles in the time interval under analysis.

REFERENCES

Aki, K. and Chouet, B., 1975, Origin of coda waves: source, attenuation and scattering effect, J.Geophys.Res., 80:3322.

Aki, K., 1980a, Attenuation of shear-waves in the lithosphere, Phys. Earth Planet.Inter., 21:50.

Aki, K., 1980b, Scattering and attenuation of shear waves in the lithosphere, J.Geophys.Res., 85:6496.

Aki, K., 1981, Scattering and attenuation of high-frequency body waves (1-25 Hz) in the lithosphere, Phys.Earth Planet.Inter., 26:241.

Console, R. and Rovelli, A., 1981, Attenuation parameters for Friuli region from strong-motion accelerogram spectra, Bull.Seismol. Soc.Am., 71:1981.

Dainty, A. M., 1981, A scattering model to explain seismic Q observations in the lithosphere between 1 and 30 Hz., Geophys.Res. Lett., 8:1126.

Fedotov, S. A. and Boldyrev, S. A., 1969, Frequency dependence of the body-wave absorption in the crust and the upper mantle of the Kuril-Island chain, Izv.Akad.Sci.USSR Phys.Solid Earth, 9:17.

Hermann, R. B., 1980, Q estimates using the coda of local earthquakes, Bull.Seismol.Soc.Am., 70:447.

Rautian, T. G. and Khalturin, V. I., 1978, The use of coda for determination of the earthquakes source spectrum, Bull.Seismol. Soc.Am., 68:923.

Spencer, T. W., Sonnad, J. R., and Butler, T. M., 1982, Seismic Q - Stratigraphy or dissipation, Geophysics, 47:16.

INDEX

Folding, 28, 30, 42, 49, 182
Fourier spectral analysis, 173
Fourier transform, 15, 134
Frequency domain, 15
Fresnel zone, 136–138

Gabbro–Diabase formations, 48
Gargano strip, 84
Geoelectrical data, 177
Geoelectrical deep explorations,
    169
Geomagnetic field, 170
Geomagnetic sounding, 170
Geophone, 15, 139–140, 160,
    165–166, 183–184, 210
  maximum amplitude, 139
  numerical representation, 144
  3-component, 183, 189, 203
Geosynclinal area, 42–43
Geosynclinal basin, 42
Geosynclinal cycle, 24, 32, 38,
    50
Geosynclinal stage, 23
Geosynclinal succession, 43
Geosynclinal trough, 42, 48
Geothermal anomaly, 159
Giudicarie Line, 86, 89
Gneiss, 180
Graben, 84
Granite
  basement, 181
  intrusion, 48–49
  layer, 32, 40, 48, 97
Granitisation, 49–50
Gravity, 32, 34, 59, 87, 97
  anomalies, 61, 87
  isostatic anomalies, 32, 34
    horizontal gradient, 34
  negative gradient, 83
Greater Caucasus, 24, 26, 28, 30,
    32, 34, 36, 38, 40, 42,
    43, 49

Greater Caucasus (continued)
  isoclinal, 28
  magmatic evolution, 44
  meganticlinorium, 24, 26, 30,
    43
    axial part, 28, 32, 42

Half-Schlumberger array, 173, 174
  curve, 174
Hardeman basin, 181
Hellenides, 114, 125, 128
  orogenic polarity, 125
Herrins T-T, 101–102
High velocity layer (HVL), 102,
    107, 119, 120, 137
Hilbert transform, 134
Horst and Graben structure, 30
Huygens' principle, 138, 149
Hydrocarbon exploration, 179
Hydrophone, 183–185
Hypocenter, 58, 62, 79, 202
  depht, 58, 61, 66, 78, 84
  distance, 208
  location, 201

Insubric Line 120, 121
International Seismological
    Centre (ISC), 7
intra-geosynclinal ridge, 120
Inversion, 42, 48
  partial, 44
  stage, 49
  velocity, 61
Ionian crust, 87
Ionian foreland, 89
Ionian plate, 191
Irpinia, 209
  earthquake, 207–208
Isostatic anomalies, 32, 34, 59
    (see also Gravity)
  field, 32
Isoseismals pattern, 61

217

219